GETTING THE BUGS Out Of ORGANIC GARDENING

GETTING THE BUGS Out Of ORGANIC GARDENING

by the Staff of
Organic Gardening and Farming

Edited by Barbara B. McKillip

Photography by Tom Fegely

Rodale Press, Inc.

Book Division
Emmaus, Pennsylvania 18049

ISBN 0-87857-056-X

Library of Congress Catalogue No. 72-93101

COPYRIGHT 1973 by RODALE PRESS, INC.

All rights reserved. No part of this publication may be reproduced or transmitted in any form or by any means, electronic or mechanical, including photocopy, recording, or any information storage and retrieval system.

ORGANIC LIVING PAPERBACKS are published by
Rodale Press, Inc., Book Division
Emmaus, Pennsylvania 18049

JB-6

FIRST PRINTING

PRINTED IN THE U.S.A.
on recycled paper

CONTENTS

Introduction: Pesticides—Who Needs Them?

Many a would-be organic gardener, or gardener making the decision to go organic—to replace his store-bought, sprayed and chemically-ripened produce with naturally-grown, succulent, vitamin-filled vegetables—may be hesitating to take that final step just because he thinks he'll have a problem. "What," he asks, "am I going to do about all those pesky little creatures that always show up just in time to bore holes in the potatoes, slice out kernels from the corn, suck the juice from the tomatoes—in short, what can an organic gardener do about BUGS?"

Rodale Press recognizes that, as the number of organic gardeners and farmers in this country grows daily by acres and acres, so does the need for insect control—insect control the natural way, without the use of poisonous pesticides. Organic gardeners know the dangers of resorting to chemical insecticides. They've seen firsthand, or read or heard about, the tragic results—the destruction of birds and wildlife, the pollution of the atmosphere, the residue-contaminated fruits and vegetables—all adding up to a threat to mankind as serious as fallout.

Organic gardeners also realize the futility of having to continually make more lethal substances in order to kill the new insect generations which have built up a resistance to the older types. They recognize that insects are not just nature's accidents but a part and parcel of the earth's corporation—a very important cog in the machinery of the life cycle.

The late J. I. Rodale believed that much of the trouble we have with insects can be avoided if we follow Nature's methods in our gardening. "The more we observe her methods, the more we come to understand that the insect is Nature's censor in destroying unwanted vegetation," he said and explained, "If a plant is growing on the wrong kind of soil, the leaves may take on an off-taste. This, as well as its appearance, is recognizable by insects. There seems to be evidence that many insects prefer to feed on plant matter that has imbalances in its chemical makeup or plant matter which is mildly diseased. Do insects prefer such food? It seems that Nature has provided them to remove the unfit. *Insects are not merely an accident. They are part of a scheme of evolution wherein they serve important functions.*"

Most gardeners spend entirely too much time fretting over bugs. Much of that time is unnecessary because the same gardeners haven't learned to live with nature instead of trying to dominate it. One of the first principles the organic gardener should adopt is the knowledge that he can't kill all of the bugs. Many of the millions of insects in our country have shown surprising resistance to all kinds of environmental abuse. If a few bugs take up residence on your plants, it's no need for general alarm.

Insects can work for you. There are some 86,000 different species of insects in this country, 76,000 of which are "friendly" or beneficial to the garden. Blankets of insecticide kill the desirable insects along with the destructive ones. Maybe the bug on your plant is a friend, not a foe, so discard the "kill at any cost" philosophy. You may destroy the "good" bugs along with the "bad." Since bugs are part of nature's scheme of things, why not tolerate them? In fact, you might like to plant a little extra so that there is more than enough for you and the bugs.

Soil comes first in controlling insects. Because insects have a craving for plants that are nutritionally deficient, you can offer the best protection by providing an organically fertile soil. Look to your soil, not a spray bottle.

Control of insect pests is possible in any number of ways, without using chemicals. New ways are being devised daily —just as often as a careful observer of insects gets an idea and an organic gardener tests it and claims success. Sometimes the ideas may seem a little weird but they're all based on the idea of working *with* nature instead of fighting her. Let the skeptics scoff—the organic gardener knows that science does not have all the answers and that, so far, Nature's way has proven best!

In this book we have outlined the concept of organic insect control, listing many alternatives—some new, some tried and true—to lethal sprays, giving personal examples and explaining how bugs themselves can be helpful in eliminating other bugs. Controlling insects can be one of the gardener's main headaches. We hope the following information will, if not obliterate the pain, at least dull it considerably!

PART I

Biological Control—— The Better Way

Organic Soil—The First Step

Chapter 1

The most controversial of all organic claims is that plants grown on rich organic soil, without use of chemical fertilizers, and without being sprayed, will not be attacked by insects because insects won't like their taste. Down through the decades, organic gardeners and farmers have made that claim because they saw with their own eyes that their plants often enjoyed an almost mysterious immunity to insect attack. Neighboring growers would have to spray several times a season, while organic gardeners and farmers could keep insects well under control by using a few biological controls plus some old-fashioned hand-picking of beetles and insect egg clusters.

Organic gardeners began saying that their plants repelled bugs because of what they saw with their own eyes, not because they read about it in books. In the early days, there were only practical results to show doubters, not laboratory evidence. However, in 1950 positive evidence was found that certain kinds of insect pests could taste the difference between healthy plants and sick plants. And they preferred the sick plants! Leonard Haseman, a professor of entomology at the University of Missouri, began to publish results of experiments which showed that insects thrived on a different nutritional mix than people—hardly surprising yet revolutionary. Here are Prof. Haseman's own words:

"It is a well established fact that man and the higher animals, in order to escape the ill effects of 'hidden hunger' and nutritional deficiency diseases, generally require a well-balanced diet rich in minerals, vitamins and other factors protecting health. Such a diet calls for vegetables, fruits, grains and other plant products grown on fertile soil, and meat and dairy products from animals fed well-enriched crops. On the other hand, not all insects require or can utilize well-balanced diets of this type. Some of them, we find, thrive and reproduce better on unbalanced or inferior diets as we think of them. Such being the case, it is but natural that these types of insect pests should thrive and increase as depletion of soil fertility requires the growing of less nutritious and less bountiful crops."

You would think a challenging idea like that—turning topsy-turvy common views about why insects attack plants—would arouse tremendous interest and spur discussion and investigation. The Haseman statements, though, were greeted with loud silence by the orthodox pest control community, which was hell-bent on a search for ever more powerful pesticide poisons. Most entomologists were not interested in what bugs liked to eat, but only in what chemicals would kill insects more effectively.

Today, Prof. Haseman's ideas are far more likely to be given wide consideration. Let's read more of what he had to say back in 1950, writing in *The Organic Farmer:*

"In nature there are abundant illustrations of insect pests choosing and breeding more abundantly on weak and undernourished plants or crops and livestock. Take for instance lice on calves. It is the weak, underfed, and undernourished, rough-coated, weaned calves and not those suckling, fat, smooth-coated ones which are often eaten up with lice. Also, a weak, sickly hen in the flock will always carry most of the lice. Again it has been found that a female mosquito may produce more eggs when it feeds on the blood of one animal than when it feeds on another.

"In like manner weakened trees, due to drought, leaky gas mains or loss of roots due to excavation, will always be more heavily attacked by borers than nearby healthy trees of the same kind. Chinch bugs tend to collect and breed more heavily on corn or wheat up on the eroded slopes rather than down at the foot of the slope where the eroded soil minerals and organic matter pile up. In this case it is the high level of nitrogen in the vigorous crop at the foot of the slope that the bug is unable to take. As is well known, the chinch bug never attacks the legumes, and soybeans planted with corn may even help to protect the corn crop from attack by the pest. To prove that high levels of nitrogen in the soil and taken up by the plant will protect corn from chinch bugs, we here at the Missouri Agricultural Experiment Station have reared the pest on seedling corn plants grown on low and high levels of nitrogen, and have found that they thrive, breed better and live longer on a diet low in nitrogen. What then is more simple in dealing with this pest than to keep soil fertility high with plenty of nitrogen supplied with legume green-manures supplemented in other ways?"

Prof. Haseman is not by any means completely alone among scientists in following a natural approach to insect control, and in pointing out that insects like to eat plants that have wrong nutritional balance. The American Association for the Advancement of Science held a seminar in 1957 on biological control of plant and animal pests, and the report of that meeting contains a fact-loaded article on the very subject we are talking about. The article—"Nutrition of the Host and Reaction to Pests" by Prof. J. G. Rodriguez of the University of Kentucky—is literally loaded with evidence that the condition of the soil can make plants either more or less attractive to pests.

First, Prof. Rodriguez says there is a strong possibility pesticides actually make plants taste better to insects. There is no question, he says, that plants take up pesticide chemicals in their leaves, because there is proof that sometimes plants are actually stimulated by small applications of DDT, for example, or are stunted by larger amounts. Therefore, although spraying kills insects, it works against the grower in the long run by making the plants more attractive to insects.

Lack of water can have a similar effect. During a dry spell, says Prof. Rodriguez, some plants take up nitrates much more quickly than the drought-stricken foliage can reduce it. The imbalance can attract mites, which "develop in high numbers in orchards and fields under drought conditions." How does the organic method help in that case? Well, ample humus in the soil is one of the best protections against drought short of an irrigation system, because humus is a perfect reservoir to balance out soil moisture reserves. Humus soaks up moisture during rains, and meters it out slowly in dry spells.

A large portion of the research reported in the Rodriguez article covers insect appetite for plants that have been grown in different artificial nutrient solutions. Such hydroponic soil mediums are a far cry from the organic method, and are in fact an exact opposite of organic methods. Yet there is some usefulness to this kind of research because it shows that when plants grow in a medium that is too imbalanced, with too much or too little of certain nutrients, insect attack tends to increase. That is meaningful to us because organic methods are perfect for balancing nutrients in the soil, just as organiculture balances water reserves better than chemical methods.

Mites seem particularly sensitive to differences in plant quality caused by fertilizer imbalances, which incidentally are very common in chemical gardening or farming situations. For example, Prof. Rodriguez says that when concentrations of all the major elements were doubled in solutions used to grow tomatoes, the mite population doubled too. Haven't you known of a chemical gardener who put on twice as much fertilizer as recommended "to be sure that the plants are well fed," then saw them eaten up by pests? Almost everyone has heard of cases like that.

Putting too much of one particular nutrient on soil can also cause trouble, by creating imbalances of other nutrients. Potassium is one of the "major" nutrients for plants, and is present in all mixed chemical fertilizers in varying amounts. But too much potassium in the soil can block uptake of magnesium, a very important element that is not usually present in mixed fertilizers. Prof. Rodriguez reported that mites thrived when potassium was high and magnesium and calcium were low. In a good organic garden, that situation is not likely to happen, because potassium and magnesium are supplied as natural rock powders, whose nutrients are made available to plants slowly, over a period of years.

Of tremendous interest in the Rodriguez article is a report from Russian scientists which "indicated that the introduction of high rates of mineral fertilizer can increase the osmotic pressure of plant sap two to three times that of normal, and also increase populations of *T. urticae,*" which is a kind of mite. "This introduces the interesting possibility," says Prof. Rodriguez, "that mites not only obtain more favorable nutrients with increased fertilization, but also may have the feeding process facilitated."

Wow, Pandora's Box is open wide! Could it be that chemical fertilizers not only make plants taste better to insects, but also literally make plants thrust themselves out for insects to nibble on?

Next we move on to aphids. "The general statement can be made," says the article "Nutrition of Host and Reaction to Pests," "that aphids respond positively to increased elements, particularly nitrogen." In other words, put more nitrogen fertilizer on your soil and you will make aphids happier. Prof. Rodriguez reports that extra nitrogen pleases the bean aphid, the cabbage aphid, the cotton aphid and the pea aphid. That's a lot of aphids.

Prof. Rodriguez sums up his review article by making a few basic points. First, he says, "because of the complex interrelationships between elements in the plant, the exact mineral requirements of a particular pest are difficult if not impossible to determine." In other words, when we talk about plants having a natural ability to repel pests, we are speaking of something that is far from an exact science. The environment of plants—soil, nutrients, light, temperature and water—vary almost indescribably. One part of a field is different from another part, one hour of the day is different from the next, and one plant responds to a certain situation differently than another plant. The chemical growers' answer to this tremendous variability is to rebel against it, to fight with chemicals for a uniform, scorched-earth policy of blasting all pests out of the garden and field.

Chemical methods are therefore completely at odds with the basic logic of nature. The natural environment is *not* uniform, and is bound eventually to trip up and disappoint any grower who attempts to garden in a rigid way, following a chemical prescription written by some white-coated dude in a laboratory.

Organic gardeners and farmers, on the other hand, weave

the variability of nature into their system, and use it to the fullest. We grow a variety of different plants, rotate crops on different plots each year, and expect something a little different from our gardens every season. The variableness of plants' resistance to insects therefore is not objectionable to us, and is in fact part of the adventure and charm of gardening by nature's way.

The last recommendation made by Prof. Rodriguez is an appeal for balance: "There appears to be a good basis for maintaining a conservative fertilization program which is not overbalanced in any one element." The organic grower is able to follow that recommendation better than anyone else, because balance is at the heart of the organic method. Organic matter is the basic balancing agent in the soil, helping to dispense both moisture and nutrients as they are needed by plants. The organic fertilizers, ranging from bone meal and dried blood to phosphate rock and dolomitic lime, do not overwhelm plants with a tremendous amount of one particular nutrient. They feed the soil first, so that the earth can in turn feed plants, animals and people in a natural way.

And yes, organically-grown plants will feed some insects too, but not as many as chemically-grown plants will. For organically-grown plants *do* have built-in resistance to insect attack.

—Robert Rodale

How to Tell the Good Bugs From the Bad

Chapter 2

Although it's very reassuring to know that out of all of the different kinds of insects there are only about 10,000 regarded as real pests, and the rest are harmless and even beneficial, how can you tell which are the good guys and which are out to do you—and your garden—in?

The beneficial insects fall into two main categories: Predators and parasites. The organic gardener may be aware that these two groups exist, but he is often unable to distinguish between them. Predators are those insects which feed on several pest *individuals* (called prey) as well as on several *kinds* of prey. Parasites, on the other hand, are usually either flies or wasps that are free-living as adults and parasites in the larval stage. An individual parasite larva usually completes its development on a *single* victim, called a host insect. The remaining beneficial insects fall into the categories of pollinators and decomposers of organic wastes.

Since predators are generally larger than parasites, their role in biological control is consequently more appreciated. Praying mantids and ladybugs are certainly the most well-known predators. One ladybug will eat 40 to 50 aphids a day, and a cup or two would be enough for the average small garden. Of course, they vary their diet, too. They're gluttons when it comes to scale insects, alfalfa weevils, Colorado potato beetles, and other soft-bodied pests, feasting on both adult insects and larvae. For the gardener who'd like to follow the success of large-scale growers, homesteaders and many others now controlling insect pests with ladybugs, here are some basic suggestions:

Plan to release a good number of ladybugs into the garden as early in the season as there is enough food to sustain them. If there is a visible infestation of pests, they will start eating as soon as thawed. Keep any for later release in refrigeration (but not freezing); they will remain dormant but healthy for several weeks.

Care must be taken in "planting" ladybugs in an infested field. They should *not* be scattered like sowing grain, but carefully placed in handful lots at the base of plants, usually a handful every 20 or 30 paces. The beetle's instinct is to climb the nearest plant, hunting food on foliage. Rough handling, especially in warm weather, may excite them to seek safety in flight. In summer months it's better to "plant" them during early morning or evening when cooler temperatures lessen their flying inclination.

Biggest enemy of ladybugs is the poisonous insecticide. Like all forms of plants and soil life, the ladybird beetle just can't take lethal sprays. Don't undo the natural balance and protection they help create by subjecting them to dusts or sprays.

When the food supply dwindles you may find many or most of your ladybugs leaving to locate better eating. If a suitable site is nearby, they may often return when the pest population increases in your section. Adult females will also lay eggs in season, thereby extending your numbers. When fall comes, some may seek sheltered nooks of a house or buildings to winter through, then emerge on the first warm days of spring.

Ladybugs are garden gluttons with a voracious appetite for aphids as well as many other varieties.

Amounts to use depend on crops, climate, type and severity of insect pests. General recommendation is for one gallon to a 5-acre truck garden, proportionately less for smaller plots and flower areas. Greenhouse culture and indoor plantings may be protected by small numbers released at the right time.

Praying mantids, odd-looking relatives of the grasshopper, are especially valuable because they have very selective diets coupled with voracious appetites. Their daily menu consists entirely of destructive garden insects. Part of the mystery which shrouds the mighty mantis is its hunting ability. Besides insect eggs and scale pests, its prey includes grasshoppers, moths, aphids, beetles, grubs and wasps. There have even been reports of mature mantises holding at bay large moths, small mice and even snakes with the adept movements of the spiny but powerful forelegs that close like jackknives around their victims.

Only during August and September—months when adult insects acquire a pair of gauzy wings for flight and mating rituals—do most people see a praying mantis. The

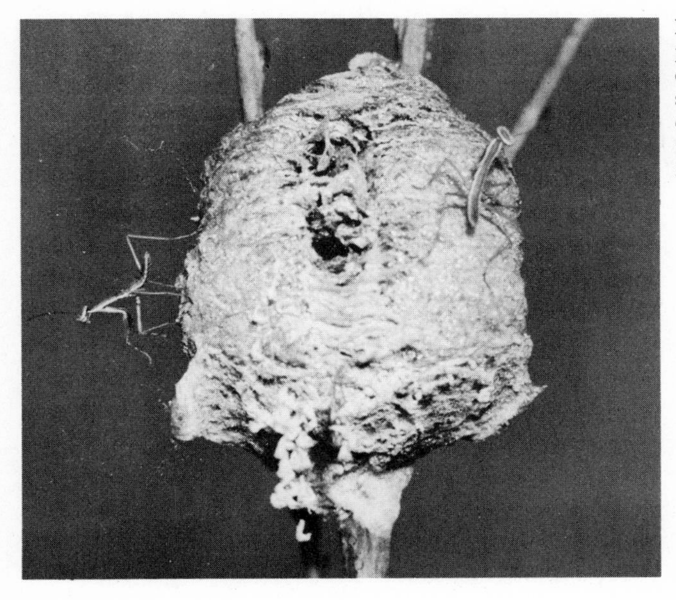

Protection of the praying mantis egg case will ensure the hatching of hundreds of the beneficial bugs in the spring.

adult females can be recognized by greatly distended abdomens, while the males have long thin profiles. And it is in late summer when the strangest part of the mantis life history unfolds. For once a female has found a mate and engaged in the mating process, she is likely to devour the smaller spouse, an instinctive provision of food for the egg formation. For some unknown reason, the more active males seem to accept this fate and don't resist. She in turn will go on and mate with five to seven more before she begins her egg laying.

Next, she selects a low-hanging branch or twig for anchorage and begins depositing layers of eggs by first secreting a fluid foam which she fashions into chambers with her abdomen. Several hundred eggs may be deposited within the case. Once finished, the female shows no further interest, as if knowing she only has 5 or 10 more days to live.

Nature provides for her eggs until spring, when they hatch to once again start the cycle. Protection of these young provides a safer and more natural way of insect control than any devised by man.

Predators that are grown and protected in one part of a garden will readily move into new plantings, following the pests they feed on. The effectiveness of predators depends on their searching capacity within the habitat. Once they find their prey, they are triggered to lay eggs; but in the absence of their prey, they will attack whatever plant-feeding pest is abundant at the moment. If faced with starvation, they can even survive on decomposing species in the organic litter, or on pollen and nectar.

The predatory group, by and large, is a buffering group, keeping all pests from increasing to dangerous levels. Predators are diverse and abundant in the environment; nearly every classification of insects contains some members which are predaceous on other insects. There are beetles, flies, wasps, lacewings, mantids, dragonflies and more. All kinds of spiders are predators, feeding entirely on insects. Even some groups of insects commonly thought of as pests contain representative species that are beneficial predators, such as certain predatory thrips and mites.

Parasite activity goes unnoticed much of the time, because of their small size and secretive habits. They should, however, be considered as no less important than predators. Parasites do not require alternate food sources in the habitat, since they feed on nectar, honeydew and other sweet secretions. In addition, they require only one host on which to complete their development. Since it is the immature parasite that kills the host, the adults can remain in the garden even when most of the pests have been destroyed; the adults then search for additional hosts on which to lay their eggs. When there are not enough pests in the garden to attract the predators, the parasites will search out the remaining insects and as a result give low-density pest control. Thus, in those gardens which are very clean of pests, it is the parasites who are responsible, since they stay on and search for hosts after the predators have left.

The first step in learning to distinguish the good bugs from the bad bugs is to redefine the term "pest." Contrary to the chemical industry's unenlightened philosophy that "the only good bug is a dead bug"—which creates a pesticide market through fear—most "crawling things" are simply that: crawling things, which do no one any harm. Indeed, many are extremely useful to mankind. So the

Eggs laid by adult parasites feed on the "host" insect, eventually killing it.

first step in studying the insects in your garden is to realize that *a garden should not be devoid of insects; indeed, if you are to achieve successful biological control in your garden, it must be filled with insects.* And some of these insects should be so-called "pest insects." In fact, what the chemical industry labels as pests (even a single individual of a potentially harmful species) is, in reality, a valuable food source for the beneficial insects in the garden.

If you stop to think about it, it is perfectly logical that the beneficial insects should need food—just as the pests do; only their food is not plants but pest insects. It is necessary to have low populations of some of the less destructive minor pest insects in the garden, to serve as food for the beneficial insects. It is *only* when these pest insects rise to intolerable levels (damaging the plants to the detriment of the yield) that they are actually *pests.* You must keep this in mind when you set out to tell the good bugs from the bad bugs. Don't panic if you detect several "pest" insects; this does not mean that your garden is in danger; it may even be a sign that your garden is strong and healthy. Remember, it is only those *excess* numbers of pest insects that are actually harmful. This is relative, of course; in the kitchen, a single housefly may be considered to be a pest; but in the chicken yard, we must tolerate a certain number of flies, for they serve as important food sources to important predators inhabiting the manure piles.

Keep in mind that there is much insect activity in the garden that you can't even see. Many parasites are so tiny that you don't even realize they are at work—until you begin to find empty pest eggs that have been sucked dry, or that have turned black, sometimes with an exit hole. The trichogramma wasp is a good case in point. Trichogramma is a helpful insect with an appetite as big as its name. Actually microscopic in size, it is fast becoming an important parasite aiding in the control of destructive insects, much the way ladybugs and the praying mantis are commonly used. The female trichogramma lays her eggs inside the egg of a harmful insect by means of a pointed egg-laying apparatus. When these hatch, the young parasites proceed to eat the contents of the egg in which they live, causing it to become black and preventing it from producing a harmful pest.

The Trichogramma does a good job on corn earworm, corn borer, looper, fruit moth (codling), pecan nut case borer, leaf tier, cabbage worm, tomato worm—and will remain in an area as long as there are eggs and larvae on which to lay their eggs. They've been especially effective in giving growers worm-free apple harvests and in curbing several other orchard and vegetable pests.

Sources for ordering trichogramma (which come carded to enable handling of the tiny insects) and other beneficial insects mentioned are listed under WHERE TO ORDER NATURAL CONTROLS at the back of this book.

Other parasites, while visible, may look like more common insects; the tachinid parasites, for example, look much like common houseflies. Predators, although very small, are much easier to detect. They vary widely in their habits. Some may actively stalk their prey; others may live nearly stationary lives, waiting for prey to approach them, or spreading traps of webbing as spiders do. Spiders themselves feed almost entirely on insects, living off whatever species is most abundant at the moment. They constitute a great buffering predaceous group that help man subdue the plant-feeding pests, and they should be considered as beneficial to your garden. Like the parasites, predators may sometimes look like other insects, sometimes pest insects. The beneficial big-eyed bug, for example, looks much like the harmful leafhopper. The only way to differentiate between them is to observe them over a period of time.

Predators are not always adult individuals. For every adult predator, there are hundreds of immature stages of growth also at work feeding on pests. This is because survival among these natural enemies is very difficult; the good bugs have their enemies, too, and they need a good habitat or they will not survive. Many eggs must hatch to provide the few that successfully reach the adult stage. But even if they die off before they reach adulthood, the tiny first stages of predators also feed on pests, usually on tiny eggs and immature pests. For example, the small lacewing larva starts out by feeding on eggs and tiny aphids; the large larvae end up by attacking larger aphids and other prey several times their size! Many lacewing eggs must hatch so that a sufficient amount of larvae will reach adulthood. So don't judge the level of predators in your garden by the adults alone; the immature ones have their own value in controlling pests.

Still another thing to keep in mind when separating the beneficial insects from the pest insects is the fact that if you see a certain bug eating a beneficial insect (an assassin bug, for example, eating a ladybug), this does *not* automatically mean that the assassin bug is a pest. In fact, this bug is beneficial, and has probably eaten hundreds of pest insects. But if there are not enough minor pests or alternate food sources present for him, he will eat a helpful insect or two. Observe carefully; *don't judge an insect to be a pest on the basis of its consumption of one or two beneficial insects.* Make sure that your habitat in the garden is beneficial for these good bugs, and then watch them for a length of time, long enough to observe their overall habits. There are no black and white snap judgments in biological control.

How do we encourage our "good" bugs? We don't discourage them with poison sprays. When we clean up the garden in the fall, we carefully set aside any praying mantis egg masses so they will be there to work for us next season. We learn to recognize the various groups of both beneficial and pest insects.

Controlling pests by natural means is not something you can turn on like a sprayer when trouble starts, but must be built up over a period of time.

Can you completely rid your garden of harmful pests by natural means? Perhaps occasionally, but usually no. The privilege of gathering a poison-free harvest will cost you 5 percent of your crop, but organic gardeners think this is a small price to pay for the satisfaction of achieving a balanced home environment and raising food fit to eat.

The point is that without man's interference, nature will set up a balance between the predators and the prey. A few insects will remain, but the damage they do will be small. The true organic gardener must strive to allow this balance to be established in his domain. He will have to tolerate some insect damage, but if he works intelligently with nature, he can keep this to a minimum. Modern man, with his disregard for his environment, and his lack of understanding of the balance of nature, is not satisfied with 95 percent of his harvest, he wants it all. So, he sprays his garden with potent insecticides and kills everything—friend and foe and the natural balance.

Insect Diseases You Can Use

Chapter 3

If bugs in your garden are making you sick, you can turn the tables on them. The use of insect disease as a means of control is becoming more widespread as new breakthroughs in effectiveness are reported.

The milky spore disease is the oldest and perhaps most widely known means of safe control for the Japanese beetle. It is actually a bacterial organism which produces a fatal disease in the grub. Because it brings about an abnormal white coloring in the insect, it was dubbed "milky." The spore ordinarily needs only a single application as it continues and spreads itself. It can be applied to the soil at any time except when the ground is frozen or when it is windy. Only mowed or cropped areas are recommended for treatment. All the gardener does is apply a teaspoonful of the spore disease powder on his grass or sod in spots three to four feet apart and in rows the same distance apart. The beetle grub, while feeding in the soil, then takes in the bacteria spores. When the grub finally succumbs to the disease, the spores that filled the body cavity are left in the soil to be picked up by other beetle grubs. As the cycle continues, the spores increase in number, more and more grubs are killed and fewer adult beetles emerge to feed on crops.

Another spore-type pest disease is a bacterium called *Bacillus thuringiensis*, now commercially produced and available, which infects a large number of caterpillar enemies of vegetable crops such as broccoli, cauliflower, lettuce, potatoes, etc., as well as alfalfa. Tests show the bacillus highly active, infecting caterpillars when young, and retaining its capacity to control susceptible insects for at least 10 years. One of the most versatile pathogens yet found in insect research (it kills more than 110 species of harmful insects), spores of *B. thuringiensis* are also being tested for control of European corn borers, pine and spruce beetles.

Tree experts and city planners have also been alerted to results of tests by USDA and state agencies that prove the new *B. thuringiensis* formula is effective and economical for control of gypsy moths. It achieved high levels of foliage protection and worm kill. Makers of Thuricide note that the newly-approved control also does the job on other major forest pests, including the fruit tree leaf-roller, spring and fall cankerworm (inchworm), fall webworm, red humped caterpillar, tent caterpillar and California oakmoth larvae, as well as a variety of vegetable raiders.

Boll weevils and tobacco budworms, two of the worst pests of cotton, have become targets of USDA studies of biological control methods. In initial laboratory tests at the Arkansas station, Fayetteville, a virus killed 64 percent of the boll weevils, infecting both larvae and adults. Called the chilo iridescent virus, it was found to be stable, and if spring field-test results are duplicated or improved, it could make a significant contribution toward weevil control, say ARS entomologists.

And an insect virus disease called nuclear polyhedral virus is still another discovery now accessible to gardeners. It infects the highly damaging cabbage looper, corn earworm, and several related pests, and has produced healthy cauliflower, cabbage, corn and other crops while nearby untreated plants have been riddled by these pests.

The Japanese beetle menace can be successfully curbed with just one application of milky spore disease.

Probably the most spectacular new biological approach is the use of sterile insects to promote destruction of a pest. In one project, for example, sex manipulation has been put into the exterminating program against the Carribean fruit fly in Florida. Millions of the flies are captured and sterilized by irradiation. When released, these sterilized flies will mate with others still in the wild—but the union will produce only sterile eggs incapable of hatching into fruit-destroying insects. So far it's working to diminish the density of the Carribean fruit fly population.

Other discoveries include a mold compound that kills or dwarfs the pesky housefly. Isolated by Drs. Raimon L. Beard and Gerald S. Walton of the Connecticut Agricultural Experimental Station, the peptide made from a common mold (*Aspergillus flavus*) produced complete control of housefly larvae within 24 hours after exposure.

One of the new controls being extensively researched, "juvenile Hormones," which confuse an insect's hormonal system and render him sterile, has been hailed as the perfect answer to all the ecological objections to pesticides. Although they are non-toxic and break down quickly into other non-toxic substances, we have serious objections about insect hormones. They are out of step with nature. There is no way we can be assured that only the harmful insects will be destroyed. Because insects regulate each other through diversity and back-up systems and insect eggs are a part of the food chain for many insects, hormones oversimplify the system. Their use can disrupt the food chains that ultimately end with man. It is another chemical rather than natural method of insect control.

Meanwhile in Florida, a fungus and a revolutionary method of applying it have been tested against the citrus rust mite, a number-one pest of the state's citrus groves and a worldwide threat to citrus grown in humid climates. Growers in Florida alone spend an estimated $5 million per year for chemical control of citrus rust mites—which rupture cells in the rind of fruit, giving it an unattractive russeted appearance, and also sap the vigor of the trees.

Entomologists established what other scientists had suspected for a number of years, that the parasitic fungus, *Hirsutella thompsonii*, destroys citrus rust mites by invading and spreading throughout the bodies of the mites.

These discoveries all add momentum to a research trend toward biological controls as safer and more effective than continued reliance on chemical insecticides. In addition, the pest diseases are self-perpetuating, spreading infection to succeeding generations. Scientists so far have identified virus diseases of more than 250 insect pests, yet have only scratched the surface of possibilities. If man can put to use just a fraction of several thousand diseases that attack insects in the wild, he will have as great a choice of biological weapons as insecticides.

More Guardians of Your Garden

Chapter 4

There are lots of other helpers ready and willing to keep you gardening organically. Remember—bugs aren't the only things that eat bugs. In addition to all those predatory insects there are other guardians of the garden like snakes and lizards (whether you like them or not) that keep rats and mice in check in addition to decreasing the insect population. Bantams, ducks, geese, bats and shrews all have reputations for keeping garden patches and orchards pest-free. The not-so-popular mole eats a diet consisting largely of wireworms, cutworms and grubs. And who could forget our most attractive allies, the birds?

Birds, as we realize now, provide gardeners with an extremely wide range of pest control. For a direct assault on grasshoppers and mice, we have crows and sparrowhawks by day. At night, owls hunt rats and mice very effectively due to their excellent hearing and the ability to see in extremely dim light. Because of their soft feathers, owls can fly more quietly than other birds and thus give their prey no warning.

Woodpeckers are especially equipped with a harpoon-like tongue to capture wood-boring insects. The woodpecker finch of the Galapagos Island, lacking the woodpecker's tongue, has learned to use a thorn as a spear to poke insects out of crevices! Woodpeckers are regarded as being of great economic value to man in the destruction of insects that harm forest trees. These birds eat vast amounts of spruce beetles and are considered the only effective control for bark beetles in the pine areas of southern United States. The codling moth has also been effectively controlled by woodpeckers in the apple orchards of Nova Scotia and Quebec.

Starlings, with their long, grubbing bills, probably deserve most of the credit for bringing the Japanese beetle under control—at least in some areas. Nighthawks and swallows have whiskers at the sides of their mouths to help them scoop up flying insects.

The killdeer is a voracious eater of grasshoppers, beetles and caterpillars; the meadowlark destroys great quantities of cutworms, caterpillars and grasshoppers; the bobwhite keeps the potato beetle under control, and orioles consume tent caterpillars. Flickers have been known to eat as many as 2,000 ants at one meal and the yellow warbler can consume 10,000 plant lice in a day.

Many gardeners put up purple martin houses in hopes of attracting a flock of these mosquito-burners to rid their grounds of mosquitoes. While the martins will put a big dent in the mosquito population, there is one thing to remember about predators: They will not rid the world of whatever it is they eat. If they did, they would soon starve. If a flock of martins stays with you all summer, you can be sure that there were enough mosquitoes to keep them fed. On the other hand, you can also be sure that—without the martins—there would have been a great many more mosquitoes.

Most gardeners have known as they watched the aerial gymnastics of a sparrow in pursuit of a darting, diving cabbage moth that there would be a lot less worms to eat their

Male Flickers feed ants to females as a part of their courtship ritual.

cabbage and cauliflower. The sparrow is also a good ground hunter, working close to your plants and picking up crawling insects or stretching up to reach the one eating the foliage. For variety in his diet, he will comb over your newly mowed lawn, usually in those spots where determined crabgrass makes seed heads between mowings.

As we can see, birds come with different appetites. Following is a list of the insects consumed by various birds.

Ants are relished by kinglets, tanagers, wood thrush, brown creepers, nuthatches, titmice, barn swallows and

Ant eggs are a special delicacy for chickadees, kinglets, gnatcatchers, titmice, nuthatches and brown creepers.

Spiders are said to be the delight of downy woodpeckers.

Weevils, capable of doing five-hundred-million dollars worth of damage annually, are devoured by the beautiful bluebirds and yellowthroat warblers whenever they are near.

Scale, minute sucking insects, make highly prized food for the little ruby-crowned kinglets, juncos, and native American sparrows (not so-called "English sparrows").

Moths would be the food most welcomed by many birds if the birds were nearby. The scarlet tanagers, phoebes, red-eyed vireos, flycatchers, gnatcatchers, and barn swallows find moths of all kinds very palatable.

Millipedes ("thousand-legged worms") bring great joy to the large fox sparrows.

Leaf hoppers are an attraction for gnatcatchers, several warblers, and others.

Grasshoppers are eagerly sought by flycatchers, bluebirds, mockingbirds, catbirds, brown thrashers and meadowlarks, and some larger birds.

Crickets are relished by scarlet tanagers, blackbirds and grackles.

Mosquitoes are most tempting to the "least" flycatchers and chimney swifts.

Other day-flying insects are regular "bill o' fare" for the flycatchers, gnatcatchers, phoebes, kinglets, barn swallows, and others.

Snails bring great satisfaction to the appetites of downy woodpeckers.

Ground insects are "gourmet specials" for towhees and juncos.

If the birds get after your fruit trees as well as the insects, you might lose some, but you'll lose many more insects. Hanging netting over the trees will help to discourage them.

Maybe he's not as pretty as our feathered friends but the toad is one of the best insurances against insect damage you can find. Once the evening dew falls, friend toad emerges from his daytime hideout under a rock, board or other shady spot and proceeds to devour countless ants, potato bugs, beetles and plant lice. He'll also feast on wasps, spiders, moths, caterpillars and flies whenever his swift, sticky tongue can snare them. It has been estimated that a single toad will consume at least 10,000 garden pests in one summer, some 2,000 of them cutworms. Slugs and mole crickets are two more of his favorites.

Since the average life of a toad is 10 years, and once content in a spot they'll come back on a steady annual basis, you can try to keep them happy by providing a cool area or water. Wetting down the shrubbery on a hot day or keeping handy a turned-over box or 8-inch clay flower pot where they can get shade will help. (Chip out a small portion of the pot for

If living conditions are right in your garden, the toad can help to keep your plants pest-free for as long as 10 years

a doorway and set in a cool, shady spot.) So will a shallow pool—into which, incidentally, you can introduce them at mating time so that their offspring will be inclined to stay around your garden. Toads will die without water and if unable to drink it, they will lie down flat in a puddle or pool to absorb moisture through their skin.

Toads possess a certain amount of homing instinct so if yours were imported from a distance, keep them penned up for awhile so they can adjust to their new environment.

For a garden free of insect pests, encourage toads to make their homes somewhere among vegetables and flowers.

Although moles and shrews may be destructive to lawns and crops at times, they also destroy many soil-inhabiting insects like white grubs and cutworms. In some areas, mice and skunks are important in controlling caterpillars and grasshoppers. Insects are the skunk's main diet item, so they are valuable friends of the farmer and the fruit grower.

An unobtrusive but extremely effective pest-exterminator is the lowly lizard. According to leading textbooks and encyclopedias, a lizard's diet consists of flies, gnats, beetles, caterpillars, slugs and other destructive insects. Lizards are common throughout North America and the most abundant of reptiles.

Lizards are fast runners so there's no danger of accidentally killing one with the power mower. Harmless to man, the lizard will flee when approached. The only thing the reptile wants to do is be free so he can go about eating insects.

Neither will lizards ever get too abundant in your yard. They are similar to certain species of birds in that the male sets up a given territory in which to feed his family, and defends this land by fighting off other intruding lizards. This is nature's way of assuring each enough food for survival. Lizards lay eggs once or twice each summer. When the one-inch baby lizards grow up, they move on to other territories.

If you'd like to try lizards in your yard for natural insect control, they can easily be caught by using a loop on the end of a fishing pole. Tie one end of a long piece of string to the tip of the pole and run the other end through the ferrule loops back to the handle. Now grasp the tip of the pole and pull the string frontward through the tip ferrule until a small loop is formed. The size and tension of this loop can be adjusted by pulling on the string near the handle of the pole.

Lizards that are impossible to catch with the bare hands are easily caught by standing at pole length and slipping the loop around their necks. When the loop is positioned, pull the string lightly and you have caught yourself a lizard. Don't slacken up on the line as he wiggles, nor tighten it too much either. With the lizard still snared on the end of the pole, simply hold him inside an open paper or cloth bag and release the tension on the line. He will wiggle out of the loop and drop into the bag.

Make sure the bag is in a place where it won't blow

The lowly lizard wants nothing more than to be allowed to pursue his favorite pastime, eating insects. Unobtrusive and harmless, he is fast becoming one of the organic gardener's most popular means of bug control.

over. Use a cloth bag or old pillowcase for carrying captured lizards if possible.

If you are unable to approach within pole length, you are probably moving too quickly or making too much noise. Slow down and be quiet. Even if the lizard runs between rocks, the loop can still be placed over his neck and he can be caught.

Lizards like to sun themselves in slightly open areas. Often they can be heard rustling leaves under bushes as you walk in the country. The pole and loop make catching these bush-dwelling lizards easy.

If you live in the city and would like lizards in your yard for bug control, they can be purchased at some pet stores. Many of the larger shops sell horned toads (not really a toad but a ground lizard) and chameleons (a good tree lizard).

When you walk in your garden and hear the rustling of dry leaves under your shrubs, you can be rest assured that a skirmishing lizard is working hard all day long keeping the pest population down to where nature intended.

Except for needing water, the box turtle, another lover of all things insect, requires no care at all. A box turtle's diet consists mainly of bugs, grubs and slugs. He's pretty fond of cutworms, too. Of course, the turtle does like various fruits and vegetables as well but what these pets take

Don't begrudge the turtle an occasional lettuce leaf. This hard-working guardian lives mainly on insects and requires only some water from the gardener.

from the garden in the way of a lettuce leaf, a fallen apple or a few blueberries or beans shouldn't worry you. It's little enough to pay for such hard-working hired hands. In fact, most of their fruits and vegetables could come from whatever's been added most recently to the compost heap so you could make it a point to leave a few imperfect tomatoes or a watermelon rind there within their reach. With the onset of cold weather, turtles dig down into the compost heap to hibernate for the winter.

One Community's Pest-Control Program

Chapter 5

If you were a stranger visiting the suburban towns of Arlington Heights, Palatine and Mount Prospect in northern Illinois, you might have been startled by the number of people studying bushes or lower branches of trees around their homes one spring. At first glance, you might think an amazing number of people in the area had suddenly become botanically-minded.

On closer scrutiny, you'd have discovered that they were observing the small, parchment-colored egg cases of the praying mantis, an insect nature endowed with a hearty appetite for other insects and a particular taste for mosquitoes. The egg cases, which had been fastened to bushes and trees, were being scrutinized to see if the mantises had hatched.

Why all this interest in a lowly insect? Simply that the citizens of these just-outside-of-Chicago suburbs, like citizens all over the country, are becoming more and more aware of the dangers inherent to the use of "hard" pesticides to control insects. And, in particular around here, concerned about the steady practice of using dangerous chemicals for mosquito fogging in suburban areas.

As a result, they formed a new but determined organization with the down-to-earth name of Pollution and Environmental Problems (PEP for short). Its purpose: "to try to return nature to the way it used to be." Along with other pollution troubles, this new group tackled the problem of combating, without the use of chemical spraying, the insect that seems to plague and irritate suburbanites the most—the lowly, bloodthirsty mosquito.

The organization went into action by not only recommending the use of praying mantis as a biological control for mosquitoes to replace chemical spraying, but they also set up the machinery to publicize and supply mantis egg cases for this purpose.

As explained by Mrs. Clayton Brown, chairman of PEP and one of its most concerned workers, "Supplying praying mantis egg cases is just one of the steps our group is taking trying to help bring back natural predators as a control means for mosquitoes."

Publicity for the campaign got underway with a booth set up by the organization in one of the area's busiest and largest shopping centers. Information and leaflets were handed out on all aspects of environmental conservation, and in particular the need to stop mosquito-abatement spraying. Orders were also taken for mantis egg cases.

Newspaper coverage of the booth's activities included feature articles—especially after alert reporters discovered the unexpected interest by the public in the mantis egg cases. Over 300 orders were taken by members manning the booth, and this total climbed to over 4,000 additional phone orders as a result of articles in the newspaper outlining the program and giving the number to call to place orders for the egg cases.

Not all newspaper coverage was favorable however. An editor of the local paper took the organization to task in an editorial condemning them for, as he put it, "tampering with nature . . . and disturbing the natural balance." In effect, this editor believed that they were breaking a cardinal rule of ecology by introducing praying mantises. The group's answer to this criticism was made by its chairman, who put into a letter the following statement: "You are correct: You don't tamper with nature. You don't disturb the natural balance. The weakness in your logic centers around the fact that man has already disturbed the natural balance, so much so that we have created a habitat that has resulted in an increased mosquito and rat population.

"Let's put it this way. Man has been using his intelligence to control 'pest' insects. However, instead of controlling the insect, man has succeeded in killing off the predators that helped control the 'pest.' When insecticide poison is poured over the land it does not distinguish between the harmful and beneficial insects."

One big obstacle to overcome in PEP's fight against

chemical spraying for mosquitoes has been the Northwest Mosquito Abatement District (NMAD), an agency formed in 1927 as a result of the action of the Illinois State Legislature.

Not only PEP, but other organizations and individuals are expressing doubt as to the effectiveness of chemical spraying to decimate the mosquito population, realizing that the relief obtained by spraying is only temporary.

Experts like Roland Eisenbeis, conservationist for the Cook County Forest Preserve District, have also expressed strong doubts. "Mosquito spraying has been practiced for about 30 years now, and what evidence is there that it has done any good?" he observes. "We've killed off, as a result, other insects that eat mosquitoes, like dragonflies as well as birds."

According to Eisenbeis, what is done locally such as spraying, is not the controlling factor, as mosquitoes can be carried by the wind and come from as far away as 20 and 30 miles. Mr. Eisenbeis has refused in past years to allow the District to spray the forest preserves with DDT. He realized then—as only a few farsighted individuals did—the disaster that could be caused by upsetting the ecological and biological balance of nature with pesticides.

"We've gone too far in our insatiable search for comfort," conservationist Eisenbeis states. "We're willing to sacrifice the whole landscape in trying to make the outdoors a big, sterile room."

PEP's 10-Point Pest Control Program

Endeavoring to go beyond mosquito control in doing their part in trying to return nature to the way it used to be, PEP is recommending the following non-toxic methods of overall insect control without the use of pesticides:

1. Choose disease- and insect-resistant plant varieties as much as possible.

2. Plant a mixture of trees, shrubs or garden plants instead of one variety. It will reduce chances of an insect outbreak.

3. Choose correct planting and harvesting dates; serious insect damage can often be avoided this way.

4. Remove dead or diseased plants to reduce the source of pest populations.

5. Rotate crops to break certain insect cycles.

6. Use proper cultivation practices to control some insects. Above all, practice good garden sanitation.

7. Encourage birds by providing food, habitat and housing for them, since they eat enormous numbers of insects.

8. Destroy insects by use of traps, inoculants and mechanical means. (Often a good water bath is very effective.)

9. Use non-toxic repellent materials such as red pepper, onion juice, flour-and-buttermilk mixture, etc. Plant certain plants such as marigold or garlic among others. These materials and plants either repel insects or aid in killing them. Some plants act as a trap crop. There are natural insecticides, such as rotenone and pyrethrum which break down rapidly, but they also can harm beneficial insects.

10. Use ladybugs and praying mantises, lacewings and trichogramma wasps for biological control.

The praying mantis insect that Illinois suburbanites are introducing into their back yards to control mosquitoes has, of course, been used for years by naturalists and gardeners along with ladybugs to effect biological control of bothersome insects.

Egg cases of the mantis are fastened to shrubs after the danger of sub-zero weather is over. Hatching usually takes place when the weather becomes warm enough. Once hatched, the young mantises are very small, about the size and appearance of mosquitoes without wings. At this stage they live on small sucking insects like aphids. By the end of a summer's eating of huge quantities of insects, the mantises attain an adult size of about four to five inches. They are then a match for any insect in the garden. Rather ferocious-looking, they are harmless to humans and will never harm plants. Mantises are expert at concealing themselves on any foliage to catch insects and to hide from birds. It is therefore quite difficult to find them. However, the patience of most gardeners is generally rewarded, as sooner or later they can catch a glimpse of a mantis in the garden, or better still in the late fall find newly-deposited egg cases —a sure sign of more of these helpful predators who will be working hard next summer in the garden "praying" and "preying" on the obnoxious and irritating mosquito.

The effects of PEP's program against insecticide spraying have been encouraging enough to warrant continuation of the praying mantis program this year. Letters and phone calls received by PEP exclaimed over the fact that for the first time in years the bird population in the suburban areas seemed to be on the increase. In addition birds that had not been seen in the area for years had returned. Some of those noted were Baltimore orioles, cedar waxwings, downy woodpeckers, and increased numbers of cardinals, nighthawks and robins.

One pleased Palatine couple who had placed several praying mantis egg cases around their property stated that it was the first summer they had really enjoyed relaxing in their backyard without being bothered greatly by mosquitoes. What made this testimonial even more noteworthy is the fact that a rather stagnant creek runs behind their yard.

PEP's Batting Average

To illustrate what "average citizens" can accomplish when aroused by the desecration and destruction of their environment, here's a list of what PEP accomplished in the six short months of its existence:

1. Encouraged natural control of insects by selling praying mantises.

2. Conducted three seminars on environmental conservation.

3. Sponsored a scrap aluminum can collection drive.

4. Printed and distributed lists of detergents and their phosphate contents.

5. Supported the Environmental Bill of Rights by obtaining signatures on petitions to forward to the Governor of Illinois.

6. Furnished speakers and information to area schools to enrich their "Earth Day" programs.

7. Conducted a major cleanup project along littered creeks and banks.

8. Conducted newspaper scrap collections for sale to paper stock company. Proceeds were used to purchase trees and shrubbery to beautify denuded areas along a railroad right of way.

. . . and they've just begun to fight.

—Armand Ferrara

A GUIDE TO SIGNS AND SYMPTOMS OF COMMON GARDEN PESTS

Insect injury to plants is a result of their attempts to secure food. As insects obtain their food either by sucking out the plant juices or by eating part of the leaf surface, the damage to the plant will vary in appearance.

The injury to plant tissues caused by the feeding of sucking insects is sometimes mistaken for a plant disease. Their needle-like beaks make scarcely visible openings, but the constant removal of the plant's juices soon begins to take effect. Most often, leaves become spotted in color.

Sucking insects also cause leaves to be yellowish, stippled white or gray. These insects, as well as their brownish eggs or excrement, can often be seen on the underside of foliage. Red spider can be spoted by yellowed leaves that are cobwebby or mealy underneath; whitish streaks mean thrips.

The second class of insects are those which damage plants

The caterpillar varieties attack plants by chewing the leaves, making their own individual patterns of destruction.

by chewing or eating the leaves. This group includes the common caterpillar, grasshopper and flea beetles, which eat small holes in the foliage of tomatoes, peppers and potatoes. Some flower garden attackers of the chewing variety are green caterpillars on nasturtiums, thrips on gladiolus and maggots on rosebuds.

The various chewing insects make their own patterns. Flea beetles make tiny round perforations; weevils produce rather typical angular openings; beetle larvae (grubs) "skeletonize" leaves, chewing everything but the epidermis and veins.

When leaves are curled up or cupped down, look out for aphids. Deformed leaves may be caused by cyclamen mite; blotches or tunnels by leaf miners; round or conical protrusions by aphids, midges or gall wasps.

The partial collapse and dying of a plant, termed wilt, may result from a number of causes—very often nematodes or gall wasps.

PART II

Keeping Insects Out With Plants

Companion Planting Can Make a Difference

Chapter 1

One of the oldest weapons known to man in the battle of the bugs is the use of the plants themselves as repellents. Plants that repel insects can serve the gardener in remarkable ways. Pests simply don't swarm or nibble on the strong-smelling, strong-tasting leaves of many herbs. Spotting them throughout the garden as a "pest control" can keep many troublemakers feeling uninvited.

Mixing crop plantings is doing it Nature's way—a diversification that brings about a balance in insect populations. Where man has substituted a monoculture or acre upon acre of the same crop, trouble has invariably followed.

Organic gardeners have discovered that bugs just don't care for any of the plants from the onion family or strong-scented herbs. At the University of California's entomology station at Riverside the discovery of the sharp pest-control effectiveness of garlic added impetus to the idea of protective plantings. Garlic planted among raspberry canes and grapevines keeps Japanese beetles at a healthy distance. Around roses, either garlic or chives will serve the same purpose. Antiseptic qualities in garlic prevent attacks by aphids. Press cloves into the soil close to the trunks of fruit trees, and plant bulbs at the base of any susceptible crop. There's no odor noticeable in the garden, nor any flavor change in foods grown. Garlic takes up little room, is easily grown and has many culinary uses.

Garlic has also been proven to be a potent destroyer of mosquitoes. Researchers at the University of California began investigating the plant when they noticed that its smell is similar to that of mosquito control. They claimed a 100 per cent kill when they sprayed breeding ponds with a garlic-based oil.

The pungent radish can be used to protect vine crops. Plant a ring around each hill and keep in all summer or make succession plantings.

Herbs, too, have been found to repel bugs effectively. Savory, for instance, has a reputation as the "bean herb." Young seedlings spaced at intervals in the furrow in place of a bean when that crop is planted will help to protect the vegetable. Basil helps tomatoes overcome both insects and disease. Set seedlings about a foot apart in the row alongside tomato plants. Mints such as spearmint, peppermint, apple mint, lemon or orange mint (bergamot) keep pests from the entire cabbage family—broccoli, cauliflower, Brussels sprout, etc. and, when hung in doorways or dog kennels, keeps the flies away.

Garlic has proven to be one of the most effective and easiest means of pest control, keeping Japanese beetles, aphids and even mosquitoes at a healthy distance.

Other bug-repelling herbs include tansy (one of the best), an attractive fern-like plant that seems to keep Japanese beetles off raspberries, blackberries, grapes and other cane fruits, as well as repelling flies from livestock and ants from the house; rue, a good-looking hardy perennial which has bitter blue-green leaves very offensive to pests, and helps dispel them among vegetables, flowers, shrubs, or trees; oregano, effective near vine crops; penny-royal, a strongly aromatic herb; and lavender, a soft, gray-foliaged plant with a vigorous, cool scent that grows well in rock gardens or among stones.

Members of the cucurbit family—pumpkins and squash also make effective fly repellents. Nip leaves carefully from strong-growing vines. Crush them and rub on the backs and heads of cattle.

Marigolds can be potent as well as pretty. Planted among the vegetables, their strong scent discourages most insects, especially nematodes and bean beetles.

Flowers, too, can do more than just stand there looking pretty. Strong-scented marigolds are probably the most beneficial, discouraging nematodes attacking potatoes, strawberries, roses and various bulbs, and displeasing Mexican bean beetles at the same time. Geraniums among roses and grapes drive the beetles away. Petunias planted around apple trees keep ants and aphids at a distance. Nasturtiums do a creditable job of chasing pests from vine crops like melons, squashes and cucumbers. Pyrethrums, related to the chrysanthemum, have their dried flowers used for insect-killing properties. Larkspur (all varieties) contains poisonous compounds that, if eaten, will cause digestive upsets in humans and kill cattle if they get enough of it so a little goes a long way in bringing about the death of Japanese beetles.

Other ways to foil the pests with plants include planting crops preferred by the bugs to keep them away from garden crops on which they might feed instead, if not given a preference. The Japanese beetle, for example, while happily feeding on the soybean crop, will ignore your cabbage, carrots, cauliflower, eggplant, lettuce, onions, parsley, peas, potatoes, radishes, spinach, squash, sweet potatoes, tomatoes and turnips. White geraniums and the odorless marigold attract the same bug, and thus protect nearby corn, raspberries, grapes and dozens of other crops. Mustard, planted early as a "catch crop", saves cabbage from the harlequin bug. The whole point of these lures is lost, of course, if the pests are not picked off and killed.

The old peasant practice of planting alternate rows of different crops is still one of the best insect-deterrents, so try alternate rows of sunflowers in your corn patch to defeat the armyworm, tomatoes near your asparagus to fight off the asparagus beetle or next to cabbage to defeat the cabbage butterfly, green beans near potatoes against the Colorado potato beetle (in turn, the potatoes help keep the Mexican beetle from attacking the beans) and horse-radish between the potato rows. Soybeans will also do a valiant job of shielding the corn from cinch bugs.

Timing your planting dates and using early or late-maturing varieties can often avoid the peak infestation periods of some pests. A lot depends on your locality's climate and growing season, but it's worth checking on. Squash planted as early as possible grows vines large enough to withstand borers, which lay their eggs in July. Radishes sown very early escape maggots, while later they can be used as a "trap crop" for onion maggots. In the Northeast, cabbage set out after June 1 usually avoids maggots. Main crops of carrots planted after June 1 will elude damage from the troublesome carrot rust fly larvae if harvested in early September before the pest's second brood hatches. Farmers have long been advised to wait until after the first week in October to sow winter wheat—when the Hessian fly is no longer active.

Crop rotation is another bug-stopping tool. Planting the same vegetable in the same place every year allows its insect and disease pests to build up. Vary the location each season, and avoid having members of the same family—such as melons, cucumber and squash—follow one another. Don't plant crops together which are attacked by the same enemies. Keep tomatoes, for instance, away from corn, since both are victims of the same worm. Potatoes troubled by the flea beetle shouldn't have tomatoes planted near them or in the same bed after them, as both suffer from its raids. Kohlrabi is another plant that will not be of benefit to the tomato.

Cover crops help too. Planting clover before seeding corn cuts down on white grubs. Repeated alfalfa cover crops gradually reduce wireworms, and other legumes minimize soil nematodes. Plant feeding habits also call for attention. Heavy feeders—like lettuce, corn, cabbage, tomatoes, cauliflower, squash, broccoli—should go into well-fertilized soil, then be followed by light feeders such as carrots, beets, turnips or other root crops. Plants that compete with each other for root space and light should be kept apart. Don't plant pole beans or potatoes close to sunflowers.

If we were limited to one pest repellent, it would be garlic. Luckily, many of these helpful plants would be growing anyway in a backyard garden. Don't expect miracles; but a variety of these planted where they fit in well go a long way in keeping pests to a minimum. Not all these plants are repugnant to all pests. Some are—probably garlic, tansy, rue, bee balm. How wide an area does a pest-repellent plant protect? Roughly within a three-foot radius. Size of the plant, strength of its odor would be factors.

Not Many Pests in My Patch

Chapter 2

How do I keep pests at a minimum without getting neurotic about it? The simplest and safest means is to use pest-repellent plants. These are mostly classed as herbs, strong-tasting and strong-smelling. Almost immune—wholly immune in many cases—these herbs also protect vulnerable crops in their close vicinity.

Some herbs have an affinity for specific crops. Basil, for instance, protects tomatoes almost as if giving them a wrap-around shield. Except for a defoliated branch or two in a whole year, my tomato plants simply do not have pests or disease. Yet they occupy the same space year after year, and their debris stays on the site over winter as part of the mulch, returning to the soil what had been taken from it. These cultural practices would be fatal if pests were there to breed in the refuse and multiply.

For many years I've grown about a dozen plants of basil (usually the handsome DARK OPAL variety, but sometimes a green-leaved kind) parallel to the tomato patch of three or four dozen plants. Varieties of basil range in height from one to two and a half feet, much smaller than most tomatoes, and are not appropriate to grow in the midst of a tomato jungle. But basil might also enclose a tomato planting. Both are tender annuals and like a warm, rich environment.

Another benefit is that the tomato tastes best as it grows best, accompanied by basil. The "royal herb," as the French call it, also lends a marvelous flavor to other foods—potato, rice, spaghetti and eggs—so even if it did not have the power to keep pests away, I would grow it.

Basil is famous as a fly-repellent, too. It can be potted up for the house or terrace, or grown in a border near an outdoor picnic area.

The herb savory, especially the annual summer savory, protects beans of all kinds. It's not so easy to grow as basil

—the very fine seeds may germinate poorly—and in my experience, it is not quite as effective a pest-chaser. Mexican bean beetles have been the most troublesome insect in my garden, although my beans have always borne well. However, each succeeding year, there have been fewer and fewer beetles—and this past summer, not a one was visible. Besides using savory, I go over the bean bushes regularly, wiping off any of the eggs or fuzzy yellow larvae on the underside of the leaves, then later hand-pick the few beetles that might appear. It helps, too, to make small, successive

Pest-repellent herbs are almost totally immune to insect attack and can keep other plants in their vicinity virtually pest-free as well.

sowings of quick-growing beans each in a different location.

Sage and mint—two more aromatic herbs—protect the cabbages and the related broccoli, cauliflower, kohlrabi and Brussels sprouts. As sprawling perennials, mint and sage can't be moved about readily. They need constant thinning and cutting back, however, especially the mint, so I use this surplus material as mulch in the cabbage row.

Mustard, another *Brassica* (same genus as cabbage), attracts the harlequin bug, if it's around, away from the cabbage. By using this "trap plant," I can pick off and kill the bugs. And of course mustard can also be put to use tastefully as cooked greens, like dandelions or turnip greens.

All my cabbage-family vegetables are bothered very little by pests. An occasional cabbage worm perhaps, but that's about all. Three purple cauliflowers, side by side in the garden one fall, were evidence of the theory that healthy plants do not usually attract pests. Two of them were fine, upstanding specimens with enormous purple heads. The one in the middle, stunted and crawling with aphids, had been knocked over by a dog in summer, and partly uprooted. I stuck it in again, but it never quite recovered. The healthy plants, so big they touched the sickly one, were free of aphids.

I really think garlic is the most valuable pest-repellent plant. Its well-known antibiotic qualities keep the garden "clean." Plant the cloves near roses, fruit trees, cabbages or any susceptible crop to keep away aphids. Japanese beetles practically disappeared from the red raspberries after I interplanted thickly with garlic. Some gardeners plant it with beans—a clove in each hill of pole beans, or a row of garlic between two rows of bush beans. A thin, upright plant, garlic takes up little room. You can't plant too much anywhere.

A pound of bulbs doesn't go far, so garlic may seem relatively costly. It represented one-third of my vegetable seed cost—the biggest item—but it was more than worth it. Of course, garlic multiplies—each clove planted producing a bulb of many cloves. But home-grown garlic tastes so much better that I harvest most of mine (in midsummer, when the tops yellow and fall over) for winter use. It's also quite hardy. What I leave for pest-repellent purposes lives on safely through winter. Tender shoots push up early in spring or even during a winter thaw.

Pests also avoid other *Alliums,* such as chives, Oriental garlic, and multiplier onions, plus their close relatives.

Marigolds rank close to garlic in value as a pest-repellent (although without the wide culinary use). A natural nematicide, they rid the soil of nematodes or eelworms, those minute worms that feed on roots of many plants, causing stunted growth. If a plant is ailing for no apparent cause, suspect nematodes. Crops liable to be damaged include tomatoes, eggplant, peppers, strawberries and beans.

The good effect of interplanting marigolds may not show the first year, since the factor that kills the worms is produced slowly in the roots of the plants, then gradually released into the soil. Grow marigolds, for instance, where tomatoes and eggplant are to go the next year. In fall, dig in the plant residue on the site, or leave it on as a mulch over winter.

The influence lasts several years, but keep on planting marigolds. Like garlic, they can't be too numerous. To save space, use dwarf kinds, which are just as effective.

Nematodes are also few where calendula (pot marigold), salvis (scarlet sage) or dahlias have grown. And as a general rule, nematodes are few in soils rich in organic matter.

Another natural nematicide is asparagus. It's a perennial crop that can't be moved, but susceptible tomatoes can be set out near the bed.

Asparagus itself, however, may be attacked by its own enemies, the asparagus beetles. The controls? For one, the nearby presence of tomatoes, which reciprocate their own protection by asparagus. Also, hand-pick beetles, and in late fall cut down and destroy the top foliage, left until then for the success of next year's crop.

Coriander and anise, a pair of annual herbs, have a reputation for repelling aphids—and after growing them, I find no reason to doubt that it's true. They seem immune themselves, while no pests were on plants nearby. And I can vouch wholeheartedly for the perennial herbs tansy (4 to 5 feet) and rue (1½ feet), which if anything does, keep pests at a distance. Both can also be used for seasoning, along with creating attractive shrub-like borders.

Still more pest-repellents include the perennial herbs—yarrow, lavender, the beautiful artemisias or wormwoods and santolina (lavender cotton). Except for yarrow tea, their culinary uses are almost nil nowadays. But all make excellent moth-repellents in the house. And a vegetable garden bordered up and down its paths with these herbs would be well-guarded. To get started, I have sown seed of lavender and yarrow, and bought a few plants of each of the others.

Among the flowers I use as pest-control helpers are coreopsis, cosmos, asters, chrysanthemums. Pyrethrum or painted daisy (the genus is *Chrysanthemum*) is more powerful. Feverfew (*C. parthenium*), sometimes also called pyrethrum, seems absolutely bug-proof, keeping pests from plants close by. Mine happens to grow in a border near roses, but it would do as much good near vegetables. Feverfew grows about 1½ feet tall, has yellow-green ferny foliage, masses of small white daisy-like flowers, and self-sows readily.

By using plants rather than poisons to keep pests to a minimum, the birds, pollinating bees, toads, beneficial insects like hornets (yes, they eat cabbage worms), wasps, ladybugs, dragon flies—even the beautiful black and orange spider—all can live safely in the environment, and the balance of nature is preserved.

—*Ruth Tirell*

Using Resistant Varieties

Chapter 3

The ideal way of protecting crops from insect injury, according to many authorities, is to develop resistant varieties. Although some progress has already been made in this work, most plant geneticists—with some notable exceptions—have been concerned with creating high-yielding or marketable crops under the assumption that insect problems can be handled by pesticides.

Most commercial seed companies have shown a distinct lack of interest in natural resistance and in looking for plants suited to the organic garden. One authority on insect resistance in plants, H. D. Pathak, discussing the resistance of corn to the European corn borer, recently stated that "many commercial hybrids being cultivated carry some resistance to this insect, but generally seed companies keep secret the level of corn-borer resistance these varieties carry." In all fairness to the seed companies, it may be possible that they don't have accurate information on the pest resistance of their various varieties, as their crops are routinely sprayed with pesticides. It does pay, however, to look for such terms as "immune" or "slightly susceptible" in descriptions of plant resistance when you're checking plant catalogs.

One thing is known: varieties do differ in their ability to attract or repel various kinds of insects which feed upon them. Why is it that some varieties can avoid or withstand an onslaught of insects with voracious appetites, while others, very closely related, are attacked and in some cases destroyed? Insects which damage agricultural crops are highly tuned to their hosts, with most of them thriving on a very narrow range of host plants. The biological fact of life suggests an important point of departure for the ecologically oriented gardener: *A change in the variety of a particular food plant may result in a change in the natural population of an insect feeding on that plant.* (Of course, such changes can take place both *for* and *against* the gardener.) If the plant is responsible for an increased population of pests on a sustained basis and is damaged as a result, it is said to be *susceptible*; whereas another variety which inhibits or reduces the numbers of the insect is said to be *resistant*. So, resistance to pests is a relative term, and depends upon a comparison with other varieties of the plant.

Plants develop pest resistance in a number of different ways. Throughout the course of its evolution a plant will cope in one way with one kind of insect, then use a different strategy with another kind. Their resistance has been categorized into three broad groups outlined below:

Type I Resistance: Non-Preference

In this category, the plant has certain characteristics which the insects do not like and which inhibit egg-laying or feeding, or deprive them of shelter.

Some varieties of rice, for example, are more resistant than others to the striped borer moths. The moths deposit only about 1/10th the number of eggs on the resistant varieties; perhaps the presence of a chemical repellent produced by the plant prevents the depositing of large egg masses, resulting in reduced pest populations. Some resistant varieties actually disrupt the behavior of an insect while feeding. The rice variety MUDGO blocks the final feeding stage of the brown plant hopper, apparently because the chemical which normally triggers feeding response in this insect is not present in sufficient concentrations in the resistant MUDGO.

Type II Resistance: Antibiosis

When a host plant has an active influence on the insect populations feeding on it, the relationship is known as *antibiosis*. There are both physical and chemical types of antibiosis. A good example of physical antibiosis is the relationship between the RESCUE variety of wheat and the wheat stem sawfly. The stem of the RESCUE variety is solid, which makes the sawfly larvae have great difficulty boring into the stems.

The biochemical forms of antibiosis are much less obvious and more difficut to locate. They can have an effect on the insect either by upsetting its metabolism, or by having a negative nutritional influence. Certain types of corn inbred for resistance to the European corn borer produce a toxic chemical which slows down the growth of borer larvae. Some plants' resistance occurs because the pests' essential foodstuffs are missing or inadequate in that variety. It is suspected that certain pea varieties which are not attacked by pea aphids are immune because they are deficient in the amino acids needed by the aphids to survive.

Type III Resistance: Tolerance Resistance

Tolerant plants are those which can host large populations of insects without suffering much damage. The same population would severely damage or kill susceptible varieties. *Tolerance resistance* is unlike other forms of resistance. *Non-preference* and *antibiosis* tend to reduce the populations of destructive insects, whereas *tolerance* actually increases their numbers beyond those found even on susceptible hosts. Tolerant plants support insect populations which might attack susceptible varieties of the same crop—in a neighbor's garden, for example. We should try to identify *tolerant* varieties. However, of the three broad categories of resistance, *tolerance* is the least preferable in agriculture.

Progress in developing insect-resistance vegetables is reported steadily from the research stations. In 1963, tomato varieties and breeding lines were reported that could resist leaf miners. In recent years, tomatoes have been reported that resist spider mites, potato aphids, tobacco flea beetles and whiteflies. It was noted that the resistance reported to spider mites and potato aphids is in horticulturally desirable lines that possess multiple disease resistance. This means two-for-one benefits for the gardeners who work with tomatoes that can handle both insects and disease.

Resistance to eight insect species has been reported among the crucifers—cabbage, broccoli, cauliflowers, etc. Insects studied include the cabbage looper, aphid, maggot and worm. Resistance to the most destructive insect of beans, the Mexican bean beetle, has also been reported. In some cases, near immunity exists regarding the potato leafhopper and snap and lima beans.

The list of vegetable varieties resistant to one or another disease or pest is growing all the time. But the problem of developing a suitable resistant variety is not as simple as merely finding a breed that doesn't appeal to the bugs. For example, entomologists at the University of Nebraska discovered a type of sweetclover that resists the onslaughts of the sweetclover weevil, but it is not suitable for commercial use. Now they have the problem of somehow transferring the resistance to commercially important sweetclover varieties. At the New York Experiment Station plant breeders developed Thaxter as a lima bean variety that was resistant to downy mildew. A problem arose when a new strain of downy mildew appeared to which Thaxter was not resistant. Plant breeders went back to work to develop a new variety with resistance to both strains of downy mildew.

Superior disease resistance in a type of wild oat and a lucky break in genetics have increased the prospect of oat varieties resistant to all known races of crown rust. Efforts to find complete resistance were intensified in 1957, when races of crown rust, rare or previously unknown in this country, first became a serious threat to U. S. oat production. According to one Department of Agriculture report, one or more of these rare races attack all varieties of cultivated oats now grown.

In the chapter on "Resistant Plants" in his book, *Chemical and Natural Control of Pests,* Dr. E. R. de Ong describes growing resistant varieties as "the ideal way of protecting crops from disease and insect injury." The discovery, breeding and selection of resistant varieties is a continuous as well as changing process. According to de Ong, environmental factors of rainfall, temperature, fertility, planting dates and soil reactions all influence the degree of resistance shown to a crop disease. Terms used to show degree of plant resistance are: slightly susceptible; moderately susceptible; extremely susceptible; and immune. Dr. de Ong writes:

"A plant may escape infection by maturing before the season of disease infection occurs. Thus, early maturing varieties of potatoes may escape late blight disease unless they are planted in mid or late summer. Disease endurance may be due to a vigorous growth that permits a plant to mature despite disease attack. Hardier structure or drought resistance may also be a factor enabling a plant to resist infection. . . . (According to Winkard,) 'True resistance to disease . . . depends on some structural or physiological characters of the plant that prevent successful invasion of the parasite.' It is a function of the genes, which are located on, and transmitted by, the chromosomes and give the hereditary characteristics of the plant. True resistance is a natural development by evolution which may be further developed by the plant-breeder."

According to *Crops and Soils Magazine* a balanced diet is important in the insect world, too. Plant parasites and cutworms, for example, die when the proteins in their normal diet are fed them in the wrong proportions.

Man is attempting to capitalize on this insect weakness. Scientists are seeking to control insect pests by changing their diet, even if it means changing the chemical nature of plants on which they feed.

R. Kasting of the Canada Department of Agriculture's research station at Lethbridge says that studies are in progress there on development of a plant deficient in certain amino acids (protein constituents) that are essential to the insect's survival. It has already been found that by plant breeding and cultural methods the balance of amino acids can be changed. But it is not yet known whether sufficient imbalance can be imparted to prevent insects from thriving.

The pale western cutworm, for example, must obtain about 10 amino acids from its food. Researchers would like the insect to continue feeding on the deficient plant until it starves to death—if such a deficient plant can be developed without jeopardizing its commercial value.

A LIST OF INSECT-DETERRENT HERBS AND PLANTS

Asters	Most insects
Basil	Repels flies and mosquitoes
Borage	Deters tomato worm—improves growth and flavor of tomatoes
Calendula	Most insects
Catnip	Deters flea beetle
Celery	White cabbage butterfly
Chrysanthemums	Deters most insects
Dead nettle	Deters potato bug—improves growth and flavor of potatoes
Eggplant	Deters Colorado potato beetle
Flax	Deters potato bug
Garlic	Deters Japanese beetle, other insects & blight
Geranium	Most insects
Horseradish	Plant at corners of potato patch to deter potato bug
Henbit	General insect repellent
Hyssop	Deters cabbage moth
Marigolds	The workhorse of the pest deterrents. Plant throughout garden to discourage Mexican bean beetles, nematodes & other insects
Mint	Deters white cabbage moth and ants
Mole Plant	Deters moles and mice if planted here & there
Nasturtium	Deters aphids, squash bugs, striped pumpkin beetles
Onion family	Deters most pests
Petunia	Protects beans
Pot Marigold	Deters asparagus beetle, tomato worms & general garden pests
Peppermint	Planted among cabbages, it repels the white cabbage butterfly
Raddish	Especially deters cucumber beetle

Rosemary	Deters cabbage moth, bean beetles & carrot fly
Rue	Deters Japanese beetle
Sage	Deters cabbage moth, carrot fly
Salsify	Repels carrot fly
Southernwood	Deters cabbage moth
Summer Savory	Deters bean beetles
Tansy	Deters flying insects, Japanese beetles, striped cucumber beetles, squash bugs, ants
Tomato	Asparagus beetle
Thyme	Deters cabbage worm
Wormwood	Carrot fly, white cabbage butterfly, black flea beetle

PART III

Other Ways to Bug the Bugs

Safe and Sane Sprays

Chapter 1

If there's a real insect orgy going on in your garden and you feel that emergency measures are called for, you *can* spray—but only with the safer, natural insecticides on the market or sprays made from plant mixtures by yourself in your mixer or blender. The commercially prepared "safe" sprays are synthesized from plants or natural substances and are said to be harmless to warm-blooded animals. Of course, we still don't know enough about their effect on beneficial insects in the garden, but if you feel the emergency is severe and immediate enough, then you may decide such sprays are worthwhile.

Quassia—the safest pesticide, has the advantage of not killing ladybugs which are eating your aphids, cheapness, and sparing bees when sprayed against apple sawfly or raspberry beetle caterpillars at blossom time. It is chips of the wood of *Picrasma quassioides* which keep dry for years in a tin, and can only be ordered through a good chemist, because it is still used by District Nurses to kill nits in children's hair. Boil 4 oz. (a pound costs about 3/6) in a gallon of water for two hours, pour off the yellow liquid when cool and dilute with five parts of water for an all-round garden spray for aphids and small caterpillars. A 1-to-3 mixture will kill gooseberry sawfly caterpillars that can strip the leaves from a bush in four days.

Ryania is a powder made by grinding up the roots of a South American plant, *Ryania speciosa.* It's useful in controlling corn borers, codling moth, Oriental fruit moth, and other pests. While the use of ryania may not actually reduce the number of harmful insects present, it will protect the crops by making the pests sick enough to lose their appetites. General recommendation is to mix about one ounce of ryania in two gallons of water when using against codling moth. Ryania has little effect against warmblooded organisms.

Rotenone, sometimes called derris, is an insecticide derived from certain tropical plants, derris, cube barbasco, timbo and a few others. It is a contact and stomach poison, often mixed with pyrethrum, and is of very low toxicity to man and animals. Like pyrethrum, it can be obtained in the pure state only from pet shops and veterinarians. When purchased in commercial dusts and sprays, rotenone is often mixed with synthetic compounds that may be toxic in varying degrees. Devil's shoestring (*Tephrosia virginiana*) is the only native plant which contains rotenone. It is a common weed in the eastern and southern states, and its roots may contain as high as five per cent rotenone. Rotenone can be safely used on all crops and ornamentals. It kills many types of insects, and also certain external parasites of animals. However, it has little residual effect and the period of protection it offers is short.

B. D. (Bio-Dynamic) Tree Spray is described as a mixture of natural compounds for sealing off all tiny and larger scars on the epidermis of plants. Because of colloidal properties, the B. D. Tree Spray is said to cover trees, shrubs, berries, etc., with a protective thin film. Basic ingredient is colloidal clay compound which is then mixed with such substances as ryania, rotenone, pyrethrum and quassia.

Used properly, a three per cent miscible oil dormant spray is effective against a host of chewing and sucking insects, organic gardeners report. Aphids, red spider, thrips, mealybugs, whiteflies, pear psylla, all kinds of scale insects and mites fall before it. The eggs of codling moth, oriental fruit moth, various leaf rollers and cankerworms are destroyed.

A dormant spray is applied to orchard trees before any of the buds open. Some gardeners make it a practice to use it on all dormant trees, shrubs and evergreens every spring, but this is rarely necessary if the plants have been organically grown for a number of years. Fruit trees, however, have many enemies and dormant spraying should be a regular practice for them, along with a strict program of sanitation.

In early spring insects that hatch from eggs laid on plants

the previous fall can be readily killed because the shells of the eggs and the protective covering of hibernating scales become softer and more porous at this time. The dormant spray penetrates and makes a tight, continuous film over these, literally suffocating the organism to death.

It will, of course, form a similar film over leaves and injure them, which is why it is applied only while the trees are in a leafless state. Citrus trees, which do not shed their leaves, are given a very dilute spray, usually made with "white oils," highly refined oils that present the least chance of foliage injury.

Dormant oil sprays have a residual effect, too. An oil film covering the plant interferes with the successful establishment of any young insects that may hatch for several days after spraying.

Stock preparations of miscible oil sprays are sold by all garden supply stores, with instructions for dilution and use. You can also make your own, using a gallon of light grade oil and a pound of fish-oil soap (an emulsifier) to a half-gallon of water. These ingredients are mixed together, brought to a boil and poured back and forth from one container to another until emulsified (thoroughly blended). Since all oil emulsions tend to separate into oil and water again, the mixture should be used as soon as possible after it is prepared. Dilute it with 20 or more times its volume of water for use.

Sometimes miscible oils are combined with Bordeaux mixture, arsenate of lead or other strong chemicals, to increase their insecticidal power. These are definitely harmful to plants and dangerous to handle, and should never be used. Lime-sulphur is also employed as a dormant spray, but it is intensely poisonous, will harm the soil and plants, and is generally less effective than an oil spray.

It is difficult to apply harmful amounts of a miscible oil spray. If too much of the emulsion is applied, the excess simply runs off. A tree should for this reason be sprayed all at once, not one-half first (as when a sprayer goes down an orchard row), the other half later, after the first has dried. Almost twice as much oil would in this case be deposited where the coverages overlap, and this could conceivably cause damage on citrus trees, especially in arid areas. The drier and warmer the air, it seems, the more likelihood of damage, though probably quite slight. Always cover a tree thoroughly in one spraying.

Sometimes oil sprays are recommended for summer use, when the trees are in leaf. This is not a good practice—it can cause leaf burn and heavy leaf and fruit drop, as well as changes in the flavor of the fruit. The insects are destroyed just as effectively, or in some cases, more effectively, in the spring.

Fruit growers in the Northeast have been using "superior dormant spray oils." These are highly refined oils which are safer to use later than the regular ones, after some new growth has appeared in the buds.

This later application is said to give a higher kill of, for example, European red mites, whose eggs become increasingly more susceptible to oil as their hatching period approaches. More mite eggs are killed with a two per cent superior oil spray applied in the delayed-dormant stage than by a four per cent spray when the trees are still dormant. The delayed dormant stages of apples is said to be when about a ½-inch of leaf tissue is exposed in the blossom buds. A three per cent superior oil concentration will destroy many oil-resistant pests.

More and more organic gardeners have discovered unique ways of making safe insecticides at home. Most of these involve mixtures of different vegetables, sometimes with spices, into solutions for natural sprays. When they find a plant that is not bothered by the pest they're after, they run these through the meat grinder, food chopper, or blender, save the liquid and residues and add equal amounts of water. They then use the ground-material soakings to spray or sprinkle the plants they want to protect. The soakings apparently contain the organic substances that keep the plant from being bothered.

Farmwives, who tend kitchen gardens outside the back door, more than likely hit upon most of the home spray concoctions listed below:

Shallots or green onions put through a food chopper, mixed with an equal amount of water, then strained through cheesecloth and sprayed over the rose bushes will chase the aphids. Onion sprays also protect tomato plants from aphids and fruitworms. Apply after every rainfall.

Effective, home-made, natural sprays have been concocted by organic gardeners through the simple trial and error system, blending mixtures from resistant plants with strong-smelling varieties, soap, salt and spices.

A solution of chopped perennial onions and feverfew is another way to keep aphids off the roses.

Hot peppers discourage all sorts of chewing insects. Grind up several of the long pods in a meat grinder, food chopper or blender. Add an equal amount of water and a teaspoon of plain dishwashing detergent, because the mixture will cling to plant leaves better if a "wetting agent" such as detergent is added. Sometimes ground onions and a bulb of garlic are added to make an all-purpose spray. The mash is covered with water and allowed to stand several hours. Next day, the mixture is strained, and water added to make a gallon. Sprayed several times a day on roses, azaleas, chrysanthemums, beans, eggplant and tomatoes, this will keep down even heavy infestation. You can bury the mash material among the plants being treated.

A jar of pickled peppers blended until all is liquid, then strained through a piece of cloth and diluted with four parts water to one of the liquid has been found especially successful in defeating maggot infestations.

Green soap, available in any drugstore, mixed with water in the ratio of one cup to three gallons respectively, will cause tree caterpillars to fall to the ground or die in the tree, leaving them easy prey for birds and other predators. Two successive daily sprayings completely clears the trees of those that survived the first.

Dissolve a half cake of Octagon soap in one gallon of hot water. Mash two garlic bulbs with four teaspoons of red pepper, add this to the water-soap mixture and spray on plants to make any bug keep its distance.

Buttermilk and wheat flour—another old standby—has been reported effective in controlling spider mites on commercial orchard crops. The mites are enveloped by the spray and killed as the spray dries on the foliage. For every hundred gallons of water, use 20 pounds of wheat flour and two quarts of buttermilk. Mix flour and buttermilk first, then add the water.

For a general spray for aphids, cut up 3 pounds of rhubarb leaves, boil for half an hour in 3 quarts of water and strain. When cool, dissolve one ounce of soapflakes in a quart of water, mix the two. It can also be made with 3 pounds of elder leaves and was used in the past as a spray for mildew on roses.

Then there are those who consider good old fashioned molasses just about the best spray material they've ever run across. They use it diluted with 50 parts water and spray it on the vegetables without worrying.

Weak salt solutions (a tablespoon in two gallons of water) have long been regarded as a good way to control cabbage worms, while an ounce of salt to a gallon of water is used to clear greenhouses of spider mites.

The value of herbs as insect repellents has already been discussed and some gardeners find an herb brew indispensable. The following recipe has proven especially potent against mealybugs:

4 parts American wormwood
4 parts desert sage
2 parts erigeron
2 parts eucalyptus
2 parts silky wormwood
1 part yerba santa
1 part wild buckwheat

Put a tablespoon of the mixed herbs into a pint of hot water and let them simmer for about 15 minutes. After cooling and straining, spray on plants.

Most of great-grandpa's garden gimmicks, invented before the advent of the commercial insecticide, are still as effective today. They included grinding mums, calendulas, marigolds or painted daisies and mixing with water for combatting the asparagus beetle; spraying with elder leaf mixture for thrips and with lime water or soap mixed with fish oil for the destruction of aphids.

Surprisingly, one of the solutions to the what-to-spray problem is as simple as the turn of a faucet. Organic gardeners have found that bugs don't like to be sprayed with water, particularly if the stream is a forceful one with large droplets. Neither do insects bother plants under a mister. Most of them can stand some rain, but few can stand being wet all the time.

By using juices squeezed from succulent plants, mixtures from protective flowers, herb teas and other natural controls instead of poisonous sprays all the beneficial members of your garden family can live safely in an environment where the balance of nature is preserved.

Tricks and Traps

Chapter 2

Traps could be your solution to insect control, just as long as they don't get rid of the good guys, too. For this reason, it's always best to try some specialized treatments before resorting to the old outdoor shaded light bulb with the kerosene-filled pan underneath.

The effect of light and color on the insects' life cycles and mechanisms has been a source of fascinating study for agricultural experiment stations, who've discovered that the pests' own individual life-styles can be used to defeat them. Some entomologists have found that blue and blue-white lights attract the most insects; yellow is less attractive, and red seems to be least visible.

Aluminum foil, a somewhat surprising ally in the battle on bugs, continues to show interesting and at times unexpected results. Initial investigation at Connecticut and other university experimental stations indicated that crops like squash, cucumbers and cantaloupes mulched with strips of aluminum foil were able to repel disease-carrying aphids and returned increased yields over unprotected plants.

Aphids can be guided to or away from host plants—depending on what color or light is reflected from the background of the plant. It has been determined that yellow attracts these insects while the blue sky reflected from aluminum foil mulch drives them away. They just keep going when they're flying over aluminum. To keep the garden clear of these insects, yellow should be used to lead the pests into traps and clear aluminum around the plants you want to protect.

The foil should be punched with holes and laid in long strips. As the plants grow through the punched holes, their foliage eventually covers the aluminum and reduces its repellency. By then, however, the protected plants have had a tremendous growth advantage over untreated ones.

Other trials showed comparable gains in lettuce, cabbage,

broccoli and several flower plantings. One test with squash at Beltsville, Md., produced a yield of 8,017 pounds per acre in the aluminum-treated plot, as opposed to 1,462 pounds in the untreated one.

Now, more recent research by the U. S. Dept. of Agriculture and the cooperating Anaconda Aluminum Co. has uncovered the fact that foil not only repels aphids but attracts some predators and beneficial pollinators. Dr. Daniel O. Wolfenbarger, entomologist at the University of Florida's Sub-Tropical Experimental Station at Homestead, noted that the honeybee population in areas treated with aluminum foil was much greater than in sections treated by other materials and in non-treated areas. Since honeybees are a source of pollination, this discovery was good news. Dr. William D. Moore, retired USDA pathologist, praised results obtained in the Florida experiments. "The big surprise," Dr. Moore said, "to most of us who have been interested in the problem of virus diseases is the advent of the use of aluminum foil in the control of aphids. Even a 50 percent control or a 40 percent control would, in many instances, spell the difference between success and failure in farming."

Some of those little buggers who only sneak out at night to do their damage may get a big surprise when their particular gardener invests in a black light trap. In a four-year test at the Agricultural Research Service in Oxford, N.C., it was shown that the tobacco hornworm population could be suppressed by a combination of black light traps and cultural practices. In another ARS experiment, black light lamps were used to lure cabbage loopers into contact with compounds that sterilized the males.

One Eastern gardener, troubled by a serious attack of chrysanthemum midge which nearly destroyed all of her 1,100 mums tried the black trap in mid-July and was amazed at the results. She found that she could open her windows or sit outside without being bothered at all by moths or mosquitoes and that the trap captured from a pint to a quart of flying insects every night. Moreover, she found no bees or other beneficial insects that she could recognize in the trap. A study of the harmful ones captured showed a surprising number, especially moths, day-flyers as well as night-flying ones.

Among the pests a black-light trap helps to curtail are the 36 different Sphinx moths, including the ones producing the sweet potato, tomato and tobacco hornworms—three of the gardner's worst enemies. Others trapped readily are the Nocturid moths, night-flyers who spawn such troublesome insects as the armyworm, cutworm, corn earworm, peach borer, gypsy moth, codling moth and Oriental fruit moth.

Baits, such as sugar, geranium oil, sassafras oil or decaying meat can be used with the black-light trap to increase the number of pests controlled. The trap can be put into use in early spring and kept in operation until after freezing weather.

Actually a form of ultra-violet light, the black-light traps are much more effective when the moon is down than when it is bright and full. There are also daylight-blue lamps which will capture the corn borer and cutworm. Light traps such as these are available to home gardeners within reasonable cost and operate at very low expense.

In contrast, homemade traps can be as simple as boards on the ground near plants, a furrow around a field or a lantern over a tub of water. Directions for an effective one suggested for the elusive earwig are as follows: Take four pieces of bamboo a foot in length, each piece open at both ends, and tie with nylon yarn into a bundle at both ends. Paint them a light green, and when dry, put under bushes, against fences and any place where earwigs are likely to gather; leave them there for a few days. Early one morning, shake earwigs out of holes into bucket of hot water or kerosene.

Earwigs can also be trapped by rolled up newspapers or jute sacking bundled in the main crotch of the tree. The traps must then be collected and burned or placed in some hot water or other more convenient killing container.

Traps baited with sassafras oil are a good control for the codling moth. A solid bait for codling moth may be made by filling a small ice-cream cup two-thirds full of sawdust, stirring into it a teaspoonful of sassafras oil and a table-spoonful of glacial acetic acid, and then adding enough liquid glue to saturate the sawdust mixture thoroughly. When the cup is dry, after a day or two, suspend it in a mason jar partly filled with water.

A molasses and water mixture can also be used. Use a piece of wire fashioned into a loop to hold a paper cup and a hook to put over a limb, then hang each cup so it is about 8 inches below the branch. Into it put a mix of one part black-strap molasses and 9 parts water, filled to the half-way mark. Use 3 or 4 for the average tree. Japanese beetles can be caught in traps filled with geranium oil.

A few well-placed traps will do much to rid the garden of grasshoppers. Half fill two-quart mason jars with 10 percent solution of molasses, and put several where the grasshoppers are at their worst.

A solution to the apple maggot problem for the small plot gardener could be bait traps similar to those used for codling moths. There are two formulas: one part blackstrap molasses, nine parts water and one cake of yeast per gallon of mixture; or substitute one part diamalt for the blackstrap molasses in the mixture. For 10 traps, add one quart of molasses or diamalt to nine quarts of water. Dissolve the yeast in water, add to the mixture and stir well. Set the mixture aside for 48 hours, or until fermentation is complete. Fill wide-mouth quart jars with a plate of ¼-inch mesh hardware cloth in place of the regular canning plate. Secure

Homemade traps of bamboo, newspapers or jute sacking have proved effective against the elusive earwig.

a pulley, ring or screw eye to a branch of the tree and run a stout cord through it in such a way that it will clear the branches when raising or lowering.

Some traps, like the hopperdozer, are designed to catch the insects that hop or jump when disturbed. It is merely a long, narrow, shallow trough of boards or metal and mounted on runners that can be drawn across a field to catch grasshoppers. A vertical shield, about three feet high, at the back of the trough is filled partly with water. Sometimes enough kerosene is added to cover the water with a thin film. The grasshoppers fly up to avoid the hopperdozer, strike the vertical shield, and fall into the kerosene-coated water. Up to 8 bushels of grasshoppers an acre have been caught with the machines. Sometimes the back and sides are coated with a sticky material. Such a device catches many leafhoppers when it is run over clover and alfalfa fields. A modified hopperdozer, merely a sticky shield or a box with the inner walls coated with a sticky material can be used for flea beetles in vegetable gardens.

Crawling insects are easy to trap, especially with traps made with a barrier. When chinch bugs, armyworms, wingless May beetles, and such are moving from one field to another they can be halted by deep, dusty-sided furrows plowed across their path. The loose dirt keeps them from escaping. The insects will fall into post holes dug at intervals along the bottom of the furrow. They can be destroyed with kerosene or crushed with a heavy stick. Irrigation ditches sometimes prevent movement of crickets from range land to irrigated fields. Sticky bands can be used to bar the progress of crawling insects (the fall and spring cankerworms on apple, elm, and other fruit and shade trees, and white-fringed beetle on pecan and other trees, and climbing cutworms on a number of host crops).

Small insects that fly or are carried by wind often are caught by a simple trap made by coating with a sticky material a piece of paper, board, wire screening, or the inside of an open-faced box, cylinder or cone. Sticky flypaper is an example. The traps may be hung on standards in yards, fields or orchards, in trees or mounted on a moving vehicle.

The color of the coated material may affect the number of insects caught when the trap is stationary.

The attractiveness of favored foods to insects can be utilized to trap many pests. Small paraffin-lined pill boxes or cans baited with sugar solutions or syrups, bacon rind, fat or meat attract ants. The ants can be destroyed by dropping the container in boiling water. Wire cages baited with a solution of one part blackstrap molasses in three parts of water, milk, or fruit waste have long been used to catch flies. The standard screened cages are 12 to 18 inches in diameter and about 24 inches in height. They have an open-end, screen-wire cone inside that reaches nearly to the top. They are set on 1-inch legs over a shallow pan containing the bait. Similar fly traps baited with meat in water have reduced blow fly populations in large areas in Texas.

Fermenting solutions and aromatic and miscellaneous chemicals attract a wide variety of insects to traps. The trap for exposing the baits may be a glass jar, stew pan, tin can, or pail; a sticky-coated baffle or support may also be used. The lure may be a simple fermenting sugar or malt solution; an aromatic element such as geraniol, methyl eugenol, oil of sassafras, or terpinyl acetate; and protein material, such as powdered egg albumen, dried yeast powder, or casein; pine tar oil, linseed oil soap, or household ammonia; or a specialized material, such as the sex attractant, identified as gyptol, that is obtained from the terminal segments of female gypsy moths.

Traps which electrocute or snare flying pests in spinning or vacuum-suction devices have also been marketed for garden and patio use and several are efficient for the sort they are designed to help curb. A home-made moth mangler designed by Paul Stevens of Oak Glen, California, consists of a 15-watt black-light bulb and two rotors turned by a fractional horsepower electric motor. The light attracts insects to the point where they can be slugged by the whirling rotor blades. Advises Stevens: "The rotor's blades must be rigid enough to kill, but not heavy enough to create a visible barrier which frightens insects."

Plastic netting, too, has come into its own as a way to

stop cabbage-family insects and others. Protective canopies of the lightweight polyethylene netting are easy to stretch over vegetable rows. Because it keeps the egg-laying parent moths of the cabbage worm away from tender young plants, gardeners report it successful in helping them to perfect crops of broccoli and cabbage without dusts or sprays. The netting comes in 100-foot-long rolls that are 7 feet wide, with either one-inch or 7/16-inch openings, similar to chicken wire. Flexible to handle, a roll costs approximately $10, will cover two 40- to 45-foot rows.

Wood ashes spread in a ring around plants and then soaked act as a deterrent to the cutworm—a particularly villainous character who chews into nearly every garden crop, usually biting off more than he can eat. Cucumber beetles, squash borers, red spiders, potato bugs and slugs also back off from the wood ash circle. Granite dust, lime, potash and phosphate rock are used in similar fashion by many gardeners.

Another long-time knack for preventing cutworm damage is banding—the use of stiff cardboard or paper collars around the stems of young plants. Push these about one inch into the soil and allow them to protrude two inches above, with a half-inch clearance around the stems. Once the plants outgrow the protective collars, they're past the critical stage and husky enough to thwart invaders on their own.

Tar-paper circles fitted around plants are another means of deterring insects from emerging seedlings. Tanglefoot, a sticky compound applied to tree trunks either in wallpaper-like strips or by pressurized spray-on containers, helps keep pests that overwinter on the ground from crawling up into the trees when they hatch in spring. Lettuce, spinach or cabbage leaves or slices of raw potato spread along the vegetable rows overnight will collect slugs, snails, cutworms, and grubs—easily gathered and disposed of the next day. Lengths of board can be employed the same way, the insects being trapped under them.

Of course, the beer-baiting idea for catching slugs and snails has become famous. All that's needed is a shallow pan, with about two inches of stale beer in it, set out where these pests are causing plant damage. Unable to resist the evils of drink, the slugs either drown or get so tipsy they can't escape.

Other ways to beat the slugs include making a barrier of coarse, dry sand, using sawdust as a mulch (adding cottonseed meal to insure against loss of nitrogen as the sawdust decomposes), sprinkling cabbage leaves with table salt which is also effective against the cabbage worm, lining garden paths with coal ashes, the texture of which is discouraging to slugs, and inverting a large cabbage leaf on the ground overnight as a trap. A good time to sprinkle powdered rock phosphate is in the spring when you're having slug trouble. Scatter with a free hand on a calm, windless day. Most of the things you can do to combat slugs will enrich your soil, too.

To trap the female cankerworm, who causes all the trouble by crawling up tree trunks and laying eggs, a protective sticky band around the trunk at the start of the spring and fall seasons will help to eliminate the pest as well as cause any survivors to move to a more favorable location. Care must be taken that the sticky bands are fresh and that there are no gaps or bridges over which the insects

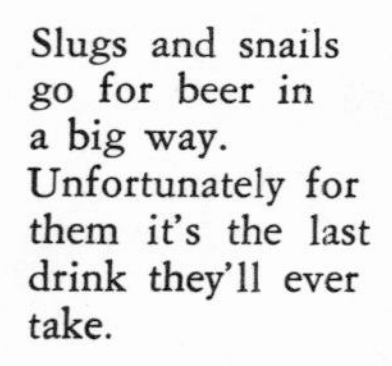
Slugs and snails go for beer in a big way. Unfortunately for them it's the last drink they'll ever take.

can cross. One banding material is Tree Tanglefoot. Cans of this adhesive compound in liquid form are available for easy spray-on application.

We may soon see more use of electronic devices for insect control. The idea of using sound as a weapon against bugs seemed inconceivable only a few years ago. Then Canada's Entomology Research Institute for Biological Control in Belleville discovered that extremely high sound vibrations could kill mosquito larvae.

Later they found that high-frequency sound resembling bat cries can help stop night-flying (tympanate) moths. Teamed with ultraviolet light traps, the sound caught four times as many moths as the light trap used alone.

A continuous beam of bat-like calls can alter the egg-laying and mating patterns of some pests, repel them from egg-laying sites. In a corn field test, one scientist used tripods with ultrasonic transducers mounted on them and succeeded in reducing an infestation of European corn borer by more than 50 per cent.

Early tests with some electronic devices also show potential. One remote-control electronic process is said to introduce vibrations into the treated plant and in so doing, it is possible to disturb insect life to the point of starvation and set up false electric atmospheres about plants which confuse insects.

Meanwhile, the small-plot gardener discovers daily such homely remedies as bacon grease and boiling water for ridding his garden of pests like yellow jackets, earwigs and ants.

Home-Tested Remedies from Our Readers

Chapter 3

Not all the ideas for nontoxic pest controls come from scientists. Over the years, organic gardening enthusiasts have done a good share of experimenting and come up with striking discoveries of their own. When they find something that really works for them they want to share it with other organic gardeners so they write jubilant articles or letters to Organic Gardening and Farming magazine like the ones we have excerpted here:

"To keep the cutworms off new plants, I cut two-inch-square bands from half-gallon wax milk containers and placed one around each new plant. These bands do a great job foiling the little pests. Garlic, chives, onions, and marigolds do a great job discouraging insects when interplanted in the garden among the vegetables."

"I read in one of my OGF issues where chives planted among roses would keep bugs away from them. Well, last year I was going to try it, but I couldn't find a chive plant or seeds anywhere. So I thought I'd try onions. I just planted a few onion sets by each rose bush. No bugs all summer. In fact, my roses were the best they have ever been."

"When a young man living in Kansas, my father used to plant a few rows of sorghum or cane between the wheat field and the corn field. A few weeks before wheat harvest, he planted the cane. Before starting to cut the wheat, he cut the cane and let it lie on the ground. Then the chinch bugs in the wheat would go into the rows of cane, eat the sap of the cane, and be killed—thus protecting the young corn."

"Last spring I set out eggplants in several spots in my garden. One was under my peach tree where I had three plants. Two died, and one was so stunted it barely remained alive until I got ready to plant my fall garden. I dug a hole and placed in it sand, gravel, river silt, barnyard fertilizer and topsoil, then transplanted it.

After giving the eggplant all the water it would take, I left it for two days while it was going through transplant shock. When I did look at it, I was surprised to see that its leaves had put on a pure-green look and it was growing at such a rate I was really glad. Next day I went out to water it and found three holes where some kind of bug had cut them. I had just gathered my hot Jalapino red peppers, so I got my pharmacist mortar and pestle and ground them into a pulp. Then I placed them flat on the ground around the stalk, taking my hands and rubbing underneath each leaf with the juice. With the ground peppers lying on the ground around the stalk, the bug didn't have a chance.

Results of my eggplant? I raised to full ripeness five of the best eggplants without another hole coming in the leaves. And the plant stayed greener than most early eggplants I had in spring planting. To my surprise, they were tastier than any I had ever raised before in my life."

"The garden gadget that did the most for my plants this past summer was aluminum foil. It started as a means of identifying a bug-chewed plant in the garden for later treating. I carried a dozen or so squares of foil around in my pocket, fitted one around each such abused plant that I found. I planned to come back later with a mixture of herbs, onion and garlic with which I sprayed these victims. But it became increasingly apparent that the foil was itself antipest treatment—and I found after an interval of just two or three days that new leaves were unchewed and the plants were doing just fine. Why? I think the pests are antifoil. I have never found one of them under it, although there are sometimes hundreds under other types of mulch. These bright squares "draw" the sun, concentrating light and heat under the plant evidently to the distaste of the predators.

Those foil-treated plants—the bug-infested ones to begin with—became the largest in the row and were the first to bloom and bear. A few vegetables, among them corn, cucumbers, eggplant and summer squash, bore so heavily that I plan fewer rows of them and more foil next summer."

Long before the U.S. Department of Agriculture's announcement that fresh or stale beer is "significantly superior to other compounds used to control slugs," thousands of OGF readers who had read Ruth Stout's articles had tried it out and knew the same truth. Incidentally, not only slugs, but the bothersome snails are curbed by simply putting out saucers or shallow pie pans of beer.

"We slugged our snails accidentally," writes a reader from Napa, Calif.

"One pleasant afternoon in April, Don, a jolly Falstaffian type, came to visit my husband. I was busy in the house but the men and our Irish terrier, Kelly, sat on the patio enjoying the delightful day. Kelly is a real member of the family and will eat anything from apples to zucchini. Neither my husband nor I care for beer, and Kelly was unacquainted with the beverage, but Don had brought some for himself and Kelly watched avidly as Don drank. The dog's imploring look was too much for kind-hearted Don, so he tracked me to the kitchen, begged a little foil pan, and filled it with beer. He placed it on the patio floor for the dog; Kelly sniffed but, for once, turned down 'people' food. The pan remained on the floor, forgotten.

"When I went outside the next morning I couldn't believe my eyes. The little pan of beer was full of slugs and snails we'd been having so much trouble with—and more were zeroing in on the target as fast as they could. Ridiculous and unbelievable? Maybe. But who was I to knock it? I called my husband and together we watched the slugs and snails making their slow torturous ways up over the side of the pan, then sinking into happy oblivion.

"We decided to experiment and see if anything so simple and safe would eradicate the pests. Various-sized containers, each holding about one and one-half inches of beer were placed around the yard. Some we sank to ground level, others we placed in likely looking spots under a shrub or in the ground cover. We even put some in the open on our cement walks. By evening every container held a respectable number of deceased slugs and snails, and the next morning several of the traps had to be emptied to make room for new victims."

Sometimes, just the smell of the stuff is enough, as one reader found:

"To control snails I have tried beer in shallow dishes. It was good thinking, but very expensive, as I do have a large garden.

I have a friend who is employed in the greenhouse in back of my property who discovered a less expensive way to trap them.

The greenhouses are very long. One day he was working in about the center of the house, drinking a bottle of beer. He didn't leave any beer in the bottle, only the odor. Then he placed the empty bottle in the flower bed and—lo and behold—when he went to pick up the botle to throw it away, there were 78 snails in it. So you see, just the smell of beer is enough. Next year I will gather empty bottles and just put in enough to give them the odor, then place them in my garden."

Another gardener had hardier slugs that just got drunk but didn't die. She solved that problem this way:

"We read your article on using beer for slug control. Ours liked the beer, but didn't die. We just had a garden full of drunk slugs! One day my husband said, "Try flour." I asked why and he said, "Since they crawl in the dampness and are damp, maybe flour will gum them up." Guess what? It worked! We took a sifter and some flour out and sifted flour all over the tomatoes and wherever else needed. What a joy to go out and pick something with no slimy slug on it."

A self-styled "hippie" from Berkeley has a different approach to the slug and snail problem:

"We consider ourselves part of Robert Rodale's 'new peasantry'—longhaired college graduates who, along with our 'radical' neighbors, have devoted much of our daily labors to the preservation and cultivation of wild life and green growing things *within* the city. Just in case you don't know it already, OGF has become required reading for much of the Berkeley 'hip' community.

"Recently I discovered an incredibly effective method of controlling the appetites of snails and slugs which can be used by anyone living in coastal areas.

"Snails and slugs are abundant in the San Francisco bay area—fortunately, so are beaches. Recently my wife and I took a trip to the nearest sandy beach where I collected several five gallon cans full of wet sand from the water's edge. Knowing of the snail's aversion to salt and abrasive surfaces, I laid out an eight to ten inch border of this sand around my garden plots. Eureka! The snails and slugs have avoided my vegetables ever since."

From Emmett, Kansas, came an idea for literally sticking fruit-tree moths, the sort that bring on wormy fruit. This gardener says she took small buckets, put about 1 cup molasses and 1½ cups water in them, adding a little sugar. "Stir up and hang these buckets in your fruit trees. You have to watch. It catches oodles of millers, the moths which lay the eggs that make the fruit wormy. You have to strain out the millers or make new solutions. This did the trick for our daughter—who canned 108 quarts of peaches from one tree, and also found she could sell her apples."

And another reader up in Canada at London, Ontario wanted us to know about her "earwig trap," which she says is what folks used 50 or more years ago. "Earwigs," she writes, "with their great fondness for chrysanthemums, dahlias, and hollyhocks are one of the worst pests of the flower garden. But they may be easily trapped in a bit of dry moss placed in a small flower pot inverted on a stake. If this trap is examined every morning and evening during June and July and the earwigs destroyed, the pest will be kept down. Pieces of hollow stems make good traps."

Our own Ruth Stout gives this simple solution to the cabbage worm problem:

"Twice a season, or possibly three times, I go down my cabbage-family row and sprinkle a little salt from a shaker on each plant. I do it right after a rain or when there's dew on the plants, and it should be done, the first time, when the plants are young, then once or twice more if there's any indication of their being eaten. This has been my procedure for several years, during which time I haven't seen one cabbage worm."

And to prevent the cabbage butterfly from laying its eggs in the first place, another reader writes:

"Cabbage butterflies do annoy Brussels sprouts. Last year I found a solution to those pests that was so easy and effective that I'll use it again this summer. I simply cut inexpensive nylon net into large squares and fitted one around each plant with paper clips. The net is crisp and stands up around the plant, doesn't cling to it even after it has been wet. I'm sure that an enterprising cabbage butterfly could lay its eggs through that slight barrier if it tried, but I've noticed that a small thing will usually stop a predator. It's also effective in preventing aphids from collecting in large numbers on the stalks in fall. This net bonnet has no effect at all on the growing plant. Of course I adjust it from time to time to keep it from strangling the stalk."

If you're tired of brown worm trails spoiling your carrots here are two ways to avoid them:

"Mix your package of carrot seed with one cup of fresh *unused* coffee grounds. Plant the coffee with your seeds. It percolates enough coffee odor during the growing season to fool the nosiest of carrot flies. It won't flavor the carrots as naphthalene and creosote do, and it's nonpoisonous."

"When the new carrots are two to three inches high, you take 1 quart 'stock' molasses (crude black molasses) to 3 gallons of water and pour over the rows of carrots. Repeat if necessary. I do this only once, never have any worms, and I have even had my carrots in the same row for 3 years."

Clover mites take a powder with this home-discovered answer from Denver:

"Ordinary talcum powder sprinkled on windowsills, at doorways, around baseboards, and around the outside foundation of your home will rid the place of pesky clover mites. Any brand, any price talcum will do.

"Denver is alive with clover mites—new home areas, older homes, and around apartment areas where dirt is brought in as a fill or for lawns often brings in a new horde. I have had drapes completely ruined from the stains. Comes a hot day and they are everywhere, even on your clothing.

"In the past we squirted thin oil from an oil can around foundations and doorways, but it was so messy. I am certain your readers will appreciate the simple method of using talcum powder."

And still the home-style alternatives to commercial insecticides keep coming, from Kansas to Wisconsin, California to New Jersey—for every pest problem from potato bugs to cantaloupe beetles:

"I read in a recent issue about volunteer horse-radish plant repelling potato bugs. I haven't any plant, but I used prepared horse-radish, which I mixed in water and applied with a window-cleaner sprayer. I haven't seen any potato bugs since. When you have the first opportunity, I would like other gardeners to know."

"This is the first year we haven't had our cabbage-family crops ruined by the white cabbage moth. A friend advised us to put salt in each hill of soil when the plants are set out. We did this, plus watering each plant with a dilute solution of our special mixture whenever I noticed the white moths hovering over the garden. It has also been very helpful when the squash and cucumber plants were almost ruined by the spotted cucumber beetle. After my 'treatment,' the beetles moved elsewhere.

"I use the blender when whipping up my concoction—

and even the dogs leave the kitchen! I cannot give specific amounts of each ingredient; let each individual vary his own.

"Ingredients used: fresh spearmint, green onion tops, garlic, horse-radish root and leaves, hot red peppers (dried, were all we had on hand in June), peppercorns and water.

"Blend all this in the mixer, then pour into a gallon jug. Add about one cup of inexpensive liquid detergent. Make as many batches as you like. When you want to use this mixture on your plants, use about ½ cup insect repellent mixture to about one quart of water. I have an old dipper that I use to give each plant a good dose.

"Up to now, the benefits have been many. I never like to use poison sprays, as ours is strictly an organic garden. Do hope this information will prove helpful to fellow gardeners."

"For 5 years I grew cantaloupes that were vigorous in plant and bountiful in fruit, only to see them die just before time for the melons to be ripe. I read somewhere in OGF that radishes planted among the melons would discourage a wilt-carrying insect.

"In every hill of melons I planted several radishes. Result: last year, no wilt, lots of melons. Tried it again this year with complete absence of wilt and plenty of delicious 5-pound melons. Apparently nature has a remedy for most ills if only we can discover those remedies."

"I have found a new way to keep cutworms from cutting vegetables in the garden. I did it by taking the stems of wild onions and tying them around my vegetables. The cutworms leave them alone. This works on flowers too, especially hollyhocks.

I would like to pass this idea along to other readers."

"I tried something new last fall. My dahlias were beautiful until the yellow or black cantaloupe beetles found them, but I made quick work of the pests. I took 3 tablespoons of ground red pepper and poured boiling hot water on it, let it set awhile, then strained it through cheesecloth into the sprayer and diluted with water. A half hour later there were no bugs—all dead on the ground. We were bug-free for the rest of the fall."

"For control of greater and lesser peach tree borers, wrap the trunk with heavy newspaper from the scaffold branches down to below the ground line. This will prevent winter injury to the tree from blowing frozen snow particles. Close the top paper with tree Tanglefoot (a sticky banding compound to stop crawling insects) and also the bottom. With this procedure, I never had a greater borer in the base of a peach tree. On the upper side of the tree (for lesser borers) use a miscible oil full strength for dormant spraying, and for summer use cut the amount in half—two tablespoons of Scalecide mixed with one gallon of water once every 7 days.

"Potatoes in my garden are more prone to bug infestation than almost any other crop, probably because they need an acid soil and mine is quite alkaline. So I tried planting them in a bed of leaves and covering them with more leaves. To my gratification, it worked. Unpestered by any bugs I could see, the potatoes grew large and healthy. I decided that the leaves gave the potatoes the lower pH they require. I've observed that vegetables planted in soil that's right for them are usually let alone by bugs that would otherwise destroy them.

Yesterday's ways can help stop today's bugs. From Springfield, Ohio came the following, favorite recollections from a somewhat-over-30 gardener:

"Get a bar of old-fashioned homemade soap (like Fels Naptha) and keep it in a bucket, filling with water. If you haven't a sprayer, take an old cup and throw the soapy water on your trees. See that this also gets on your fruit tree blossoms. This likewise will take care of worms and bugs on your grapevines. . . . Repeat after every hard rain during the growing season. . . . When I have a few more insects (mosquitoes, etc.) than I want, I go to a general feed store and buy some bone meal. It is an insect repellent as well as a good fertilizer for your plants. It repels wild rabbits, too. . . . I save my old cooking oil, and on a nice warm winter day, I put in the oil some dissolved soap, maybe some garlic juice, etc., and take an old paint brush and use it up and down the trunk and nearby branches of my trees. This takes care of bugs and beetles that winter under the bark."

Finally, Illinois gardener Anita Stanley discovered the first rule of combat was *Know your enemy* and armed with that information, set out to study the habits and haunts of the bugs in her garden and how to destroy them before they destroyed the plants. What follows are some of her findings:

"Because they appear first in spring, I started with the *ants.* Sometimes their large hills cause soil around shallow-rooted plants to become fine and loose so it can't retain enough water. Often ants cause damage by transplanting aphids and scale insects from one plant to another, while feeding on their sticky secretions known as honeydew.

"Since birds, toads and lizards are natural enemies of ants, I planted a grape arbor near my garden as the ant-eating robins find them ideal nesting places. I provided a wren house nearby. To help these allies, I sprinkle red-hot pepper into the ant hills.

"*Aphids* were next on my list of pests to study since there were so many of them. Some grow wings enabling them to fly from one plant to another—when they're not riding the wind for many miles. Male aphids appear only in fall, and after mating, the females lay eggs which live over winter to hatch out in spring.

"Since aphids produce sticky secretions on trees that attract ants, they are double trouble as a sooty, black fungus may develop to ruin the tree completely. Aphids can be killed only by using something that penetrates their bodies. (The same is true of *mealy bugs.*) After much experimenting, I found that soapsuds made from naptha soap will do this. After applying suds, spray hard with plain water every day for awhile to keep any new eggs from hatching out.

"Most *beetles* are harmful as their young bore into plants or form grubs in lawns or gardens. *Blister beetles* (ones I dislike most) are tough to destroy; a container of boric acid and water set in their abiding place will help if the weather is dry. Against *Japanese beetles,* I find that making small trenches alongside the rows of vegetables and laying rags there soaked with turpentine (derived from pine) discourages reproduction.

"It's a simple job to seek out and destroy the first adults of the *Mexican bean beetle*—one-third inch long, yellow-brown with 16 black spots. I boil cedar sawdust or chips in water, let the mixture cool, and then spray all bean plants with the soluiton. After a few daily applications, I find my bean plants unmolested again.

"I use the "cedar-water" mixture against other pests, such as *potato beetles, red spider,* and *mealy bugs* (indoors you can swab house plants with alcohol).

"Cedar sawdust and cedar chips also work against some troublemakers. *Cucumber beetles* and *squash bugs* are at their worst in July and August, so I work quite a lot of cedar chips and sawdust into the ground as I plant these sections of the garden. When the plants begin to vine out, I soak newspapers with turpentine and lay them over the

Squash bugs are foiled by cedar chips and sawdust used in combination with turpentine-soaked newspapers.

ground letting the plants vine over them. Sifted cedar sawdust works well against *chiggers* in the lawn.

"*Borers* can be a constant source of irritation. They are the caterpillar stage of moths and the grub stage of beetles. Usually they work right inside the plant, and many types work just under the bark of fruit and other trees. Newly-planted trees can be protected by wrapping the trunk with heavy paper or burlap. Painting the crotches of trees and shrubs with a mixture of kerosene and lime seems good protection. Cedar sawdust and lime makes an effective paste too.

"If you're having trouble with *cutworms, wireworms,* or *grubs* try removing all mulch from your garden this fall and plowing or tilling the soil well to expose the larva. Here in central Illinois this practice will winterkill most of them.

"*Green worms* are the larvae of the white butterfly moths that attack cabbage, cauliflower, etc. The best defense I learned is to encase the young plants with nylon net. This prevents butterflies from depositing larvae that later develops into the marauding green worms.

"Abnormal growths on leaves, stems and petals, *galls*

Grasshoppers cease to be a nuisance in a garden with a prowling cat or two.

form on many types of plants. They are caused by a mite or insect laying its eggs and the young hatching out develop in the plant tissue. Whenever I see a gall on leaf or stem, I clip it, drop it into a can of kerosene, and check the spread of these destructive insect growths.

"*Grasshoppers* I didn't have to learn about, for on the farm in Iowa where my deceased husband was raised, they are sometimes so thick as to become a plague. They eat foliage and flowers of plants to the extent they can strip a corn field. Usually they are not so numerous in my garden as to do much damage. Perhaps because my neighbor's cats find them delectable tidbits. Let a couple of cats prowl your garden for a few days and the grasshoppers disappear.

"Whenever I see about one-eighth of the tips of arbor-vitae turn brown, I know the *leaf miner* is there. These are small insects that eat just under the surface of leaves so that they riddle them with light-colored spots or long, narrow tunnels. When their presence is evident, I immediately douse the plants thoroughly with naptha soapsuds. (I find it's a good idea to douse many plants with soapsuds about the time daffodils come into bloom to kill pest larvae which survived the winter.) Soapsuds, worked well under the leaves of plants, are also effective against leaf hoppers —the 'Typhoid Mary's' of the plant world carrying diseases.

"*Lacebugs* are as destructive as they are beautiful. The artist in me is fascinated by their intricate design even as I plot to exterminate them. They are about one-eighth inch long with delicate lacy wings, but they quickly ruin foliage by sucking the sap from the underside of the leaves and covering it with sticky honeydew. Look for lacebugs early in spring, especially around azaleas and rhododendrons. Several dousings with heavy soapsuds will destroy them.

"One of my worst enemies is *red spider*. I've found it on house plants, in the garden, and on my beloved cedars, pines and blue spruces. Not longer than one-sixteenth inch, they cause evergreen needles to turn a grayish brown. Spraying in spring with miscible oil (mixed with water) or cedar water will usually kill their eggs. I watch for signs of them as the weather becomes hot and dry, and then bless the friend who told me about spraying with strong coffee. It works! I sprinkle young trees with soaked coffee grounds, which cling into the tops where red spider is most harmful.

"*Spittle bugs* with their frothy mass are all too familiar. They thrive in hay fields; therefore I mulch strawberries with straw instead of hay as they also thrive on strawberries. Heavy rains tend to control them. I find that mixing cedar and lime into hay mulch discourages the emergence of young in many types of pests.

"The smallest of all pests, *thrips* are about the size of a lettuce seed and often the same color. Worst ones infest gladiolus, roses, onions and peonies. As soon as peony buds begin to show color, I douse them with a mixture of kerosene and soapy water. If rose buds fail to open or seem to rot, I douse them too.

"I have named and described here some of the common types of pests to be found in nearly every garden. Some leave something for me to eat, others are more greedy. Learning to know these pests, their breeding habits, etc., has enabled me to control them, so that I can enjoy the good things from the garden.

"There are, of course, many more that I haven't mentioned here, but these are the main ones. I learned that persistence in seeking out breeding places, finding the eggs and the larva and destroying them is the easiest and best means of control. By learning where and when to look for these—by *knowing my enemies*—I've become a much better and happier gardener."

Specific Alternatives to Insecticides

Ant

Keep ants away by banding plants with tanglefoot. That prevents insects from climbing up the plant. In home gardens, steamed bone meal has been found to discourage ants. If ants persist, a pepper spray may make them think twice. (Shellac the exterior of ruined bark or wood in nearby trees to take away their favorite habitats.) Tansy has been found to discourage ants. Plant it at the back door on house foundation to keep both ants and flies from the house. The dried leaves of tansy sprinkled about—in cellar or attic—are a harmless indoor "insecticide."

Fire ants respond to different adverse habitat changes, particularly the level of humidity in the soil. They require somewhat moist conditions to nest so that drying the area can help keep them away. Opening the nest with a shovel and pouring weed-oil into it will also deter nesting. A more or less continuous attack is needed to reduce the fire ant's numbers.

Aphid

Enrich your soil organically, as the aphid detests plants grown in organically rich soil. Grow nasturtiums, which repel aphids, between your vegetable rows and around fruit trees. Or make a trap of a small pan painted bright yellow and filled with detergent water. The aphids will become attracted to the bright yellow color and alight on the water surface, trapping themselves. You can trick aphids by placing some shiny aluminum foil around your plants so that it reflects the brilliance and heat of the sun. Aphids shy away from foil mulched plants. Encourage the aphid's natural enemy, the ladybug, who eats many times her weight in aphids.

If aphids are a serious problem, try a mixture of tobacco

water or soapsuds. Using tobacco stems (usually available from florist or nurseryman) as a base, mix a handful into a gallon of water, letting it stand for 24 hours. Dilute it to color of weak tea and syringe the foliage, making sure to hit the undersides of the leaves.

Or you can add soap to water, apply the suds to the plants, and rinse with clear water afterward. Some gardeners report success with a solution of turnip mash (crushed raw turnips) mixed with corn oil. Still another defense against aphids—peach leaf curl also—is the easy-to-make onion-water spray. (Experiment with the concentration best suited to you and your garden.)

Bagworm

Bagworms are easily removed by hand. Black light traps are also effective for catching the worms in their moth stage. Trichogramma are a natural and effective enemy.

Bean Beetle

Encourage beneficial praying mantes, which have been most effective in bean beetle control. Also, plant your heaviest crop of beans for canning and freezing early, because those plants are freer of the pest than late season ones. Don't forget to use interplanting techniques with potatoes, nasturtiums, cloves and garlic. Some gardeners have used a mixture of crushed turnips and corn oil to foil the beetle.

Cabbage Maggot

Create a strong alkaline area around the plants to deter the maggot, by placing a heaping tablespoon of wood ashes around each plant stem, mixing some soil around with the ashes and setting the plants in firmly. Protective canopies of polyethylene netting also prevent infestation by keeping insects from laying their eggs in the young plants. In addition, practice general insect control measures such as the use of sanitation, rotation and improvement of the soil condition. Interplanting might be able to beat the cabbage worm. Surround your cabbage by cole plants such as tomatoes and sage which are shunned by the cabbage butterfly, the parent of the green worm. Further interplanting techniques include the use of tansy, rosemary, sage, nasturtium, catnip and hyssop. Friendly insects like trichogramma are available from commercial sources. Avoid using poison sprays which will kill these and most helpful insects. In addition, homemade, non-toxic sprays such as pepper sprays, sour milk sprays and salt mixtures have been found effective.

Caterpillar, Tent

Perhaps the best control for caterpillar is the use of *Baccillus thuringiensis.* This non-harmful bacterial is eaten by the insects who become paralyzed and die in about 24 hours. Other control measures include the use of sticky bands so that the female worms, who are unable to fly, will not be able to crawl up trees and lay eggs there. You may wish to place burlap or shaggy bands around the trees to attract the caterpillars and trap them. Those caught can be destroyed daily. Light traps are also effective. Encourage praying mantes, birds. Trichogramma is effective against the tent caterpillar as well as the gypsy moth and cankerworm. Remove and destroy the brown egg masses from the branches of any wild cherry trees in vacant lots and other areas around your home. Every time a single egg mass is destroyed, the potential threat of 200 to 300 more tent caterpillars next spring is gone.

Chinch Bug

This little black sucking insect can cause large brown patches in your lawn, and all but destroy your sweet corn. Chinch bugs thrive on nitrogen-deficient plants, so heavy applications of compost will make your plants resistant and avoid much of the trouble. If present in your lawn, remove the soil from the spot, and replace it with one-third crushed rock, one-third sharp builder's sand, and one-third compost. If they show up in the corn patch, plant soybeans as a companion drop to shade the bases of the corn plants, making them less desirable to the highly destructive chinch bug.

Codling Moth

Place a band of corrugated paper around the main branches and trunk of affected trees. When larva spin their cocoons inside the corrugations, they can be removed and turned. Eliminate loose bark from trunks and limbs where moths like to hibernate. Trichogramma, tiny female insects, are helpful in biological control of moths. Also effective is a black-light-bulb trap which attracts and kills the moths during their summer sessions. One gardener suspends a container of molasses and water mixture in his trees to trap the moth. Ryania will discourage codling moth, and birds are effective natural controls. Woodpeckers consume more than 50 percent of the codling moth larvae during the winter. Don't discourage them by saturating everything with viperous spray-can solutions.

Corn Borer

Destroy overwintering borers by disposing of infested cornstalks. Plow or turn under the refuge or relegate it to the compost heap. Plant resistant or tolerant strains of corn —consult your county agent for the best hybrids available locally. Because moths lay their eggs on the earliest planted corn, it is generally advisable to plant as late as possible, staying within the normal growing period for your locality. Encourage parasites like the spotted lady beetle which eats the eggs of the borer on an average of almost 60 per day. Trichogramma is also a natural enemy of the corn borer.

Electric light traps work effectively to protect a corn patch against borers. The commercially-available *Bacillus thuringiensis* makes another good defensive weapon.

Corn Earworm

Fill a medicine dropper with clear mineral oil and apply it into the silk of the tip of each ear. Apply only after the silks have wilted and have begun to turn brown. Another easier control, but not as sure, is to cut the silk off close to the ear every four days.

Cucumber Beetle

Heavy mulching is a time-tested control. For every bad infestation, spray by mixing a handful of wood ashes and a handful of lime in two gallons of water. Apply to both sides of leaves. The *spotted cucumber beetle* and family are repelled by frequent "dustings" of bone meal and rock phosphate. Usually apply in early morning, since powder sticks to plants better when they're wet with dew. Watch plants closely for signs of beetle damage, and apply mixture as often as needed (possibly 2 or 3 times weekly).

Radishes, marigolds and nasturtiums offer interplanting protection.

Cutworms

Cutworms chew plants off at the ground level. They work at night and hide beneath soil or other shelter during the day. A simple device for preventing damage is to place a paper collar around the stem extending for some distance below and above the ground level. Some gardeners get the same effect from using tin cans, with "electroculture" benefits as an added plus. Toads and bantam hens are natural feeders of cutworms. Cultivate lightly around the base of the plant to dig up the culprit first. Keep down weeds and grasses on which the cutworm moth lays its eggs. Interplanting with onions has been found to be effective in many cases.

Earwigs

Traps have been an effective control for earwigs. Bantam hens feast on the earwigs and do a good job of eliminating them from the home grounds. Set out shallow tins of water which will attract earwigs so they can be destroyed.

Flea Beetles

Clean culture, weed control and removal of crop remnants will help to prevent damage from flea beetles. Control weeds both in the garden and along the margins. Since flea beetles are sometimes driven away by shade, interplant susceptible crops near shade-giving ones. Tillage right after harvest makes the soil unattractive to egg-laying females and will assist in destroying eggs already laid.

Grasshoppers

Virtually every kind of bird has a craving for grasshoppers. Some eat the eggs after scratching them from the ground. Construct birdhouses and otherwise attract birds to your garden if you experience difficulty from grasshoppers. Grasshoppers can be baited by using buckets or tubs of water and a light placed nearby. Because grasshoppers lay their eggs in soil not covered with plants, keep a good ground cover to prevent egg laying in the soil. Turn the soil in the spring to a depth of 8 inches so the eggs will not hatch. Eliminate any weeds around garden margins.

Japanese Beetles

Perhaps the most important control organism is the "milky spore disease," a bacterial organism that creates a fatal disease in the grub. The disease is caused by a germ not harmful to man and is available commercially. Since the beetles are attracted to poorly nourished trees and plants, be certain your soil is enriched by the addition of plenty of organic matter. Remove prematurely ripening or diseased fruit, an attractive dish for the beetles. Eliminate weeds and other sources of infestation like poison ivy and wild fox grape. Some gardeners get effective results by interplanting larkspur.

Leafhoppers

Leafhoppers seem to prefer open areas, so plant your crops near houses or in protected areas to avoid damage. It's also a good idea to enclose your garden plants in a cheesecloth or muslin supported on wooden frames. Pyrethrum is an effective nonpoisonous control that can be dusted on top of the plants. Keep the neighborhood clean and raked so that insects will be exposed to the weather, particularly during the autumn. Avoid planting susceptible varieties.

Maggot

Use tar paper collars around the stems to prevent the flies from laying eggs on the plants. Place plants in irregular rows so that the maggot is not able to find them easily. In the case of onions, this random technique often offers increased protection to neighboring plants because the onion smell is repulsive to many garden pests. Hot pepper spray is an easy and certain control.

Mealybug

Wash off plants with a strong stream of water or use fir tree oil. Denatured alcohol can be used on house plants and may be successful in light infestations. Cultivate or turn the soil for several weeks before planting to kill any grass or weeds which may be hosts. Also scatter bone meal to ward off fire ants which often carry individual mealybugs from plant to plant.

Mites and Red Spiders

These pests thrive in stagnate, very humid air so try to give your plants good air circulation. Remove mites from plants by spraying forcibly with plain water, being sure to hit the undersides of the leaves. (Generally, spiders washed off plants do not return.) A three per cent oil spray has also been found effective. Ladybugs are the mites biggest nemesis, so encourage their visit to your garden. An onion spray has been found effective but do not use poison sprays as they usually kill the enemies of mites and spiders but do not kill mites themselves. Pyrethium is a safe dust-type control and can be used both indoors and out.

Mosquito

Mosquitoes may be controlled by draining stagnant bodies of water or by floating on them on a thin film of oil. While this may be somewhat injurious to vegetation, it is not as dangerous as DDT or other poisons. Often rain barrels and other containers with water become mosquito breeding places. Eliminate those from the home ground. Perhaps the best control in your immediate area is to encourage birds like the purple martin, just one of which will eat 2,000 mosquitoes a day. The praying mantis is also a natural enemy. Agricultural experiment stations have had some success with using a garlic spray.

Nematode

The consistent use of compost will virtually eliminate nematodes. Avoid chemicals to exterminate them, as that will interfere with the proper functioning of beneficial soil organisms which tend to keep out all dangerous microbes and nematodes. Organic fertilization of infected plants induces the formation of roots and improves plant vigor, thus negating the harmful effects of nematodes feeding on the roots. The most practical answer to the nematode problem for the average gardener is to build up the humus content of the soil and to interplant with marigolds, especially the French or African varieties.

Peach Borer

Protect peach trees by keeping the ground beneath them perfectly clean of grass, weeds and mulch for at least a foot in all directions to discourage rodents and other animal pests. This also enables birds to get to the young borers. Swab each one of your peach trees with tanglefoot before planting. The substance will catch the moths or the worms so that they cannot penetrate the material or get into the bark. Planting garlic cloves close to the tree trunk has been found effective against the borer as has the trichogramma.

Potato Bug

A deep mulch—a consistently good practice for plant protection—is especially recommended for potatoes. The idea is to cover the potato seed with a one-foot layer of hay or straw—thereby creating a barrier for the bug but not the plant. At season's end, the mulch can be worked into the soil to build its humus content.

Another standard control is to encourage and add ladybugs, praying mantids, trichogramma, etc.—all commercially available.

Root Maggot

Repel root maggots by applications of large quantities of unleeched wood ashes or a mulch of oak leaves if available. Be sure to locate your growing area away from members of the cabbage family for at least three years. Maggots are particularly attracted to radishes, and some gardeners plant them as a trap crop to be discarded later. If infestation is heavy, test the soil and feed it. Then grow a cover crop to be turned under.

Scale

Best control is to spray infected trees early in the spring with a dormant oil emulsion spray. Ladybugs, available commercially, feast on scale insects and usually keep these pests under control.

Slug and Snail

Snails and slugs tend to be nocturnal. Take advantage of their nighttime habits by placing shingles, planks, boards or other similar material in the garden to serve as traps. Each morning destroy those which have hidden away there for the day. The bodies of snails and slugs are soft and highly sensitive to sharp objects such as sand and soil and such dry and slightly corrosive substances such as slaked lime and wood ashes. A narrow border of sharp sand or cinders around a bed or border will serve as an effective barrier against them as will a sprinkling of slaked lime or wood ashes. Many gardeners have found that setting out saucers of beer, sunk to ground level, attracts slugs by the droves so they can easily be destroyed.

Sowbug

The best control for sowbugs is prevention. Look for and eliminate hiding places in and around the home garden area. Make certain that logs, boards and other damp places are eliminated. Frogs and poultry like to feast on sowbugs, so if you're lucky enough to have some around, turn them loose on this garden villain. If not, discourage them with a light sprinkling of lime.

Squash Bug

Squash bugs can be repelled from squash and other susceptible plants by growing radishes, nasturtiums and marigolds nearby. Hand picking of either the eggs, nymph or adult stage is effective in a small garden. Because the squash bug likes the damp moist protected areas, he often hibernates under piles of boards or in buildings. By placing boards on the soil around your plants you might be able to trap him and easily destroy the bugs every morning. For an effective spray, mix a handful of wood ashes and a handful of hydrated lime in a sprinkling can of water. Let the mixture stand for a day. Use a sprayer or the can to apply whenever attacks are evident.

Tomato Hornworm

Tomato hornworms may be hand picked and killed by depositing them in a small can of kerosene. On a larger commercial scale, the tomato grower may obtain effective control from light traps, since the hornworm must pass through the moth stage in its life cycle. If the back of the hornworm is covered with a cluster of small white bodies, do not hand pick. Those are parasites which will kill the worm and live to prey on others. Trichogramma, praying mantis and a ground hot pepper dusting all prove good controls.

White Fly

In the greenhouse or for indoor gardens generally, a planting of Peruvian ground cherry (*Nicandra physalodes*) is an effective white fly repellent. So is hanging fly ribbons on a stick. Outdoors, test for phosphorus deficiency in the soil. Ladybugs are fond of white flies, too. Many gardeners remove white flies from their garden area along with dandelion heads by using a vacuum cleaner to suck them up. In small greenhouses improved air circulation by exhaust fans is also helpful.

Wireworms

Good drainage tends to reduce wireworm damage. Newly-broken sod land should not be used for the garden if other soil is available. If sod must be used, it should be thoroughly plowed or stirred once a week for several weeks in early spring. Stirring the soil exposes many of the insects and crushes others. Enriching soil with humus will also improve aeration and reduce wireworm attacks. Plant radishes and turnips as a trap crop.

Where to Order Natural Controls

INSECT BIOLOGICAL CONTROLS

TRIK-O (Trade name for Trichogramma wasps)
Gothard, Inc.
P.O. Box 370
Canutillo, Tx. 79835

(Recommended for: Flower and Vegetable gardens, Berries, Grapes, Fruit & Nut Trees and many field crops; controls apple coddling moth worm.

Vitova Insectary, Inc.
P.O. Box 475
Rialto, Ca. 92376

(Lacewings and Trichogramma wasps and fly control parasites)

Eastern Biological Control Co.
Route 5, Box 379
Jackson, N.J. 08527
(Trichogramma wasps)

INSECT DISEASE CONTROLS

"DOOM" (Milky Disease Spores Control
Japanese Beetle Grubs and other Grubs.)
Fairfax Biological Laboratory
Clinton Corners, N.Y. 12514

THURICIDE (Bacillus Thuringiensis)

(Recommended for: Lawn Moth Larvae and Caterpillar Control, Safe, non-toxic to humans and animals.)

International Mineral & Chemical Corp.
Crop Aid Products Dept.
5401 Old Orchard Road
Skokie, Ill. 60076

BIOTROL (Bacillus Thuringiensis)
Thompson-Hayward Chemical Company
P.O. Box 2383
Kansas City, Ks. 66110

PLANT-DERIVED INSECTICIDES

RYANIA
Hopkins Agricultural Chemical Company
P.O. Box 584
Madison, Wis. 53701
(Controls Coddling Moths on Apples-Corn Borers on Corn

ROTENONE AND PYRETHRUM
Garden Supply Stores & Seed Firms
(In pure state: veterinarians & pet shops)

B. D. Tree Spray (Bio-Dynamic Product)
Peter A. Escher
Threefold Farm
Spring Valley, N.Y. 10977

HERBS

Pine Hills Herb Farms
P.O. Box 144
Roswell, Ga. 30075

Nichols Garden Nursery
1190 N. Pacific Hwy.
Albany, Or. 97321

Meadowbrook Herb Garden
Rte. #138
Wyoming, R.I. 02898

Casa Yerba
P.O. Box 176
Tustin, Ca. 02827

Greene Herb Gardens
Greene, R.I. 02827

DORMANT OIL SPRAY

SCALECIDE
B. G. Pratt Co.
206 21st Avenue
Patterson, N.J. 07503

(Kills: Scale, Red Mites, Aphids.
Spray before new growth starts on fruit trees, shade trees, ornamentals)

TREE TANGLEFOOT

The Tanglefoot Company
314 Straight Ave., S. W.
Grand Rapids, Mich. 49500
(Protects trees from crawling insects)

PLASTIC NETTING

Animal Repellents
P.O. Box 168
Griffin, Ga. 30223

Apex Mills, Inc.
49 West 37th Street
New York, N.Y. 10018

Frank Coviello
1300 83rd Street
North Bergen, N.J. 07047
(Cheesecloth protects ripening fruit from birds)

LADYBUGS

Bio-Control Company
Route 2, Box 2397
Auburn, Ca. 95603

L. E. Schnoor
Rough & Ready, Ca. 95975

Montgomery Ward
618 W. Chicago Avenue
Chicago, Ill. 60610

PRAYING MANTIDS

Eastern Biological Control Co.
Route 5, Box 379
Jackson, N.J. 08527

Montgomery Ward
618 W. Chicago Avenue
Chicago, Ill. 60610

NOTE: These materials or equipment must be used according to directions for best results. Please order from the advertisers noted here or from your seed, hardware or garden supply store. Remember that some plants and predators, as well as dormant oil sprays are seasonal.

Index